SpringerBriefs in Energy

More information about this series at http://www.springer.com/series/8903

Francesca Stazi · Federica Naspi

Impact of Occupants' Behaviour on Zero-Energy Buildings

 Springer

Francesca Stazi
Department of Materials, Environmental
  Sciences and Urban Planning (SIMAU)
Università Politecnica delle Marche
Ancona
Italy

Federica Naspi
Department of Construction, Civil
  Engineering and Architecture (DICEA)
Università Politecnica delle Marche
Ancona
Italy

ISSN 2191-5520   ISSN 2191-5539   (electronic)
SpringerBriefs in Energy
ISBN 978-3-319-71866-8   ISBN 978-3-319-71867-5   (eBook)
https://doi.org/10.1007/978-3-319-71867-5

Library of Congress Control Number: 2017959541

Printed on acid-free paper

This Springer imprint is published by Springer Nature
The registered company is Springer International Publishing AG
The registered company address is: Gewerbestrasse 11, 6330 Cham, Switzerland

Contents

Acronyms

ABM	Agent-based model
AC	Air-conditioning
BAS	Building automation system
BCVTB	Building Controls Virtual Test Bed
BDI	Belief–Desire–Intention
BEPS	Building energy performance simulators
BIM	Building Information Modeling
BMS	Building management system
CFD	Computational fluid dynamics
EMS	Energy management system
EPBD	Energy Performance of Buildings Directive
FMI	Functional Mock-up Interface
FMU	Functional Mock-up Unit
GHG	Greenhouse gas
HPB	High performance building
HVAC	Heating, Ventilation and Air Conditioning
IAQ	Indoor air quality
IEA EBC	International Energy Agency Energy in Buildings and Community
IEQ	Indoor environmental quality
LOOCV	Leave-one-out cross-validation
MSE	Mean squared error
MV	Mechanically ventilated
NV	Naturally ventilated
nZEB	net Zero Energy Buildings
PIR	Passive infrared
RFID	Radio frequency identification
XML	eXtensible Markup Language
ZEB	Zero energy buildings

Chapter 1
Introduction

Users' presence and behaviours inside buildings are the main causes of energy consumptions. In fact, nowadays the construction sector is responsible for about 40% of the energy demand in EU and it is the main contributor to greenhouse gas (GHG) emissions [1]. A correct assessment of the energy request is a challenge of primary importance since it influences both the environmental and the economic aspect.

When incorrect evaluations concern low-energy buildings, additional issues are generated. Net-zero and zero energy buildings should satisfy precise requirements in terms of energy demand and indoor comfort conditions and, as a consequence, users' expectations are very high.

The IEA EBC Annex 53: Total Energy Use in Buildings [2] also includes occupants' behaviour among the six main factors that influence the energy use in buildings, together with climate, building envelope, building energy and services systems, indoor design criteria, building operation and maintenance.

The evaluation of the impact of such components on building energy performances and on the indoor environmental quality is performed using building energy performance simulators (BEPS). Since reducing energy demand and assessing adequate indoor conditions are two of the main target of the building sector, many efforts have been made to improve BEPS skills. Simulators have become very accurate in modelling and simulating deterministic features that influence building's performances (e.g. simultaneous solvers of thermal balance, electrical power and Computational Fluid Dynamics (CFD)) [3]. However, they are still lacking in representing stochastic features connected to occupants' aspects [4, 5].

The users' goal in buildings is to achieve and maintain acceptable indoor comfort conditions (i.e. thermal, visual, acoustic and indoor air quality (IAQ)). Their comfort sensations are a function of both objective and subjective aspects. The former are related to building properties (e.g. exposure) and environmental conditions, while the latter concern physiological and psychological features, peculiar for each person. Triggered by all these stimuli, occupants interact with building devices to modify the surrounding and to restore their favourite sensation.

© The Author(s) 2018
F. Stazi and F. Naspi, *Impact of Occupants' Behaviour on Zero-Energy Buildings*,
SpringerBriefs in Energy, https://doi.org/10.1007/978-3-319-71867-5_1

Such adaptations have immediate and tangible consequences on the indoor environment and on users' comfort but their effects on building energy consumptions are not of a secondary relevance. This continuous cycle is depicted in Fig. 1.1.

In the perspective of assessing the optimal indoor conditions for users' satisfactions, several comfort models have been proposed. At first the Fanger "static" model [6], which considers occupants as containers that passively undergo the building management; then the adaptive comfort model [7–9] which, conversely, regards the users as active subjects that modify the surrounding according to their preferences and needs. Despite the different attributes of these approaches, comfort models are limited to evaluating whether or not comfort conditions are matched. The advance in such perspective concerns the assessment of the behaviours the users take to adjust the environment, interacting with building systems and devices. For example, comfort models evaluating a discomfort for warm environment only suggest that the indoor temperature is too high but do not give any provisions on window opening possibility.

Moving in this direction, many researchers directed their efforts in understanding, representing and reproducing people's behaviours in buildings through the development of behavioural models [10]. A behavioural model empirically analyses and understands the interactions between the people and the environment. It is usually based on experimental data that allow identifying how the environment affects the occupants and the way the users adapt to that with adaptive actions. In this perspective, the IEA EBC Annex 66 (Definition and Simulation of Occupant Behaviour in Buildings) [11] focuses on the influence of occupants' behaviour on building energy performance. In particular, the project aims at the development of a standardised description and classification of occupants' behaviours and at the implementation of behavioural models in energy simulators.

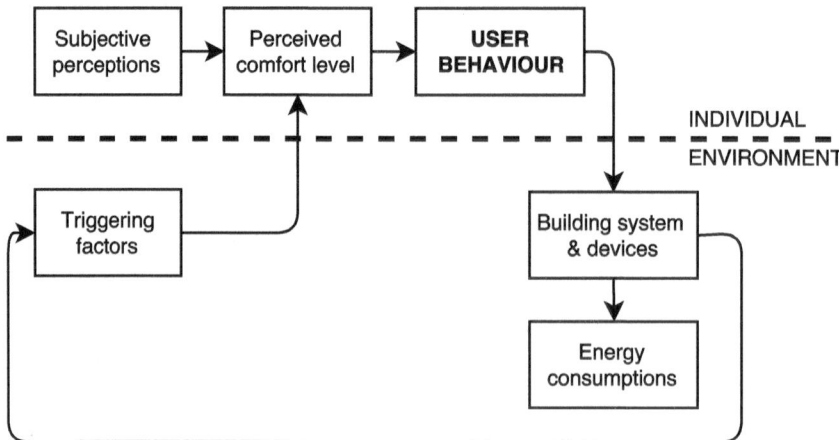

Fig. 1.1 Interaction between the users and the environment

The inclusion of human perspective has been successfully adopted also in others design contexts, as the people's safety during critical events (e.g. fires and earthquakes) [12, 13], with the aim to outperform the standard and schematic design approaches.

In fact, the correct representation and integration of the human component inside simulation software is an essential step to enhance the simulation results, reduce the gap between real and predicted energy consumptions [14–16], improve building design quality, control and operation. A behavioural approach will also have a tangible influence on the decision-making process, aiding building designers in the assessment and comparison of different strategies with more accuracy [17]. Figure 1.2 highlights the differences between a standard and a behavioural approach, underlying the effect on simulations' results. In fact, in the first case, the adoption of deterministic rules to simulate occupants' presence and behaviours could bring to incorrect evaluations at the design stage, which cause the energy performance gap. On the contrary, the accurate previsions obtained adopting the behavioural approach, should lead to a sensible reduction of the performance gap.

This book highlights the importance to include the human component in all the steps of the design process, reporting and discussing the strategies adopted until now. Furthermore, the analyses and findings presented along the book are supported by novel experimental data derived from monitoring campaigns carried out by our research group.

Hence, the first part of this book (Chaps. 2 and 3) discusses the impact of users' behaviours on building performance and indoor environment. Respectively, Chap. 2 focuses on the effects of people's actions on net-zero-energy buildings, reporting several results from literature. The target of Chap. 3 is investigating the perturbations of the indoor environment due to occupants' behaviours, also in relation to different building uses.

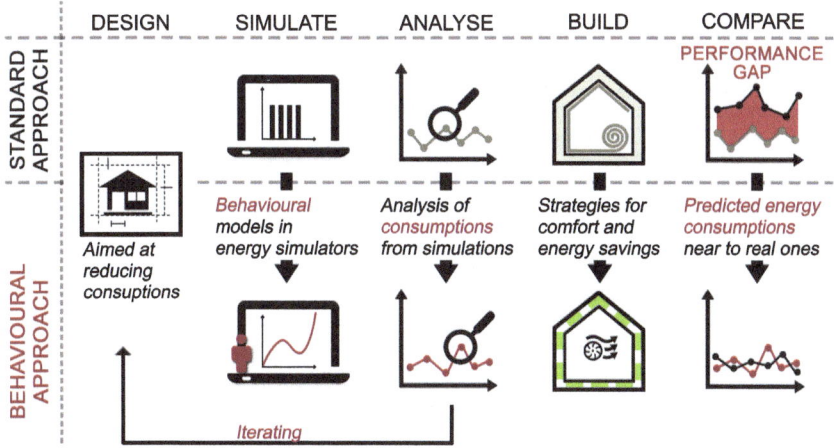

Fig. 1.2 Standard versus behavioural approach

Chapter 4 provides an overview of the factors that trigger occupants' behaviours. It analyses both objective (e.g. environmental parameters) and subjective aspects (e.g. psychological features) and highlights how the occupants' perspective changes in energy efficient buildings. Once the drivers have been assessed, Chap. 5 illustrates the principal adaptive actions that users can take in buildings. The final part of the chapter concerns the sequences of behaviours the users can take.

Chapters 6 and 7 regard the behavioural models' development. At first (Chap. 6), the techniques to acquire the necessary experimental data are provided. Then, Chap. 7 presents several approaches usually adopted to develop the models; it also discusses the methods to implement such models in BEPS and to validate them.

Finally, Chap. 8 summarises the main goals reached until now in the energy behaviour research field and outlines the future steps and challenges to face with.

Appendices A and B present an overview of the cases study cited in the book and practical applications with experimental data.

The intent of this book is to offer a better understanding of occupants' behaviours in the built environment. It also provides guidelines for designers and researchers to enhance simulation results and, as a consequence, to guide both the building design and management phases.

References

1. WBCSD (2010) Energy efficiency in buildings PPP: multi-annual roadmap and longer- term strategy, Technical Report. World Bus Counc Sustain Dev. https://doi.org/10.2777/10074
2. IEA (2012) Annex 53 task force, Final report, Total energy use in residential buildings—the modeling and simulation of occupant behavior
3. Schweiker M, Haldi F, Shukuya M, Robinson D (2012) Verification of stochastic models of window opening behaviour for residential buildings. J Build Perform Simul 5:55–74. https://doi.org/10.1080/19401493.2011.567422
4. Hong T, Taylor-Lange SC, D'Oca S et al (2016) Advances in research and applications of energy-related occupant behavior in buildings. Energy Build 116:694–702. https://doi.org/10.1016/j.enbuild.2015.11.052
5. Yan D, O'brien W, Hong T et al (2015) Occupant behavior modeling for building performance simulation: current state and future challenges. Energy Build 107:264–278. https://doi.org/10.1016/j.enbuild.2015.08.032
6. Fanger PO (1970) Thermal comfort: analysis and applications in environmental engineering. New York
7. De Dear R, Brager G, Cooper D (1997) Developing an Adaptive Model of Thermal Comfort and Preference. ASHRAE Transactions 145–167
8. Nicol JF, Humphreys MA (2002) Adaptive thermal comfort and sustainable thermal standards for buildings. Energy Build 34:563–572. https://doi.org/10.1016/S0378-7788(02)00006-3
9. de Dear RJ, Brager GS (2002) Thermal comfort in naturally ventilated buildings: revisions to ASHRAE Standard 55. Energy Build 34:549–561. https://doi.org/10.1016/S0378-7788(02)00005-1
10. Stazi F, Naspi F, D'Orazio M (2017) A literature review on driving factors and contextual events influencing occupants' behaviours in buildings. Build Environ 118:40–66. https://doi.org/10.1016/j.buildenv.2017.03.021

11. Yan D, Hong T (2014) IEA EBC Annex 66: Definition and Simulation of Occupant Behavior in Buildings.
12. Bernardini G (2017) Fire safety of historical. Buildings. https://doi.org/10.1007/978-3-319-55744-1
13. Bernardini G, D'Orazio M, Quagliarini E (2016) Towards a "behavioural design" approach for seismic risk reduction strategies of buildings and their environment. Saf Sci 86:273–294. https://doi.org/10.1016/j.ssci.2016.03.010
14. Gaetani I, Hoes PJ, Hensen JLM (2016) Occupant behavior in building energy simulation: towards a fit-for-purpose modeling strategy. Energy Build 121:188–204. https://doi.org/10.1016/j.enbuild.2016.03.038
15. Menezes AC, Cripps A, Bouchlaghem D, Buswell R (2012) Predicted vs. actual energy performance of non-domestic buildings: using post-occupancy evaluation data to reduce the performance gap. Appl Energy 97:355–364. https://doi.org/10.1016/j.apenergy.2011.11.075
16. Calì D, Osterhage T, Streblow R, Müller D (2016) Energy performance gap in refurbished German dwellings: lesson learned from a field test. Energy Build 127:1146–1158. https://doi.org/10.1016/j.enbuild.2016.05.020
17. O'Brien W, Gaetani I, Gilani S, et al (2016) International survey on current occupant modelling approaches in building performance simulation. J Build Perform Simul 1–19. https://doi.org/10.1080/19401493.2016.1243731

Chapter 2
Importance of Energy Prevision Accuracy for Zero-Energy Buildings

Abstract The energy-related attitude of occupants can cause big variations between real and predicted building energy consumptions. Such discrepancies have been found especially high in nZEBs and energy-efficient buildings. Considering that actual regulations impose the nZEB target and that in such optimised contexts users' actions have a considerable impact, a comprehensive knowledge of occupants' behaviours should be the main goal of the building sector to effectively reach the expected performance. In this perspective, this chapter firstly analyses the human-building interactions in nZEBs, by focusing on the energy implications of their attitudes. Then, nZEBs' performances are investigated and reasons for the operation gaps are highlighted. Finally, literature results from energy simulations are reviewed to evaluate the influence of different users' energy attitudes.

2.1 Introduction

Users' presence and behaviours have considerable consequences on building energy performance [1]. In fact, the actions and the management the occupants take are the main reasons for energy consumptions [2–4]. In low-energy buildings and, in particular, in nZEBs the strategies applied to the envelope and the systems in order to optimise the energy use, produce a delicate balance. In such contexts, people's direct (e.g. devices' adjustments) and indirect (e.g. thermal loads) influence assumes an increasing influence [5] and the prediction of their behaviours is even more complex [6].

Accurate forecasts of building performances are of primary importance since the building sector is unavoidably moving towards nZEB constructions in relation to both public and private buildings.

The EPBD, 2010/31/EC [7], which regulates the imminent diffusion of nZEBs, also reports the unique standardised definition of these buildings. The directive describes such construction typology as the one concerning buildings with very high energy performance, where the nearly zero or very low amount of energy required should be extensively covered by renewable sources produced on-site or

© The Author(s) 2018   7
F. Stazi and F. Naspi, *Impact of Occupants' Behaviour on Zero-Energy Buildings*,
SpringerBriefs in Energy, https://doi.org/10.1007/978-3-319-71867-5_2

nearby. Despite this definition, the EPBD neither prescribes a common approach to implementing nZEB nor describes the assessment categories in detail. For these reasons, each country has established different parameters in their own definition.

In most countries, the nZEB definition refers to maximum primary energy as one of the main indicators. In a few cases (e.g. the Netherlands), the primary energy use of the building is assessed through a non-dimensional coefficient, comparing the building's primary energy use with a "reference" building with similar character-istics. In several countries (e.g. the UK and Spain) carbon emissions are used as the main indicator, while in others (e.g. Austria) carbon emissions are used as a complementary indicator to primary energy use.

The nZEBs offer as a further step in comparison to green buildings since they are usually characterised by high-performance envelopes, ad-hoc designed systems and on-site energy production and recovery, targeting at a considerable reduction of energy consumptions.

All these features make that the human component has a greater weight in such buildings than in standard ones and, as a consequence, it usually causes larger discrepancies in the predicted error for energy consumptions [8]. While the average error in standard buildings is roughly 30% [9], studies have found differences up to 50% between design evaluations and measurements in nZEBs [10].

The following sub-sections report findings from previous studies that highlight the importance of the human factor in the specific case of nZEBs. Starting from a discussion on the interactions between the users and the building (Sect. 2.2), it is underlined how people affect building's performance referencing to results from both surveys (Sect. 2.3) and simulations reported in the literature (Sect. 2.4).

2.2 User-Building Interaction

People's reactions to discomfort can lead to opposite consequences in buildings performances. Wise and conscious users' behaviours can aid the building to per-form as well as possible, conversely, wasteful actions can cause relevant energy increases.

In fact, when people feel discomfort, they can operate three different types of adaptations: environmental, personal and psychological [11]. The first one is the most tangible and it concretely influences the indoor environment since it concerns the modification of the devices' status (e.g. windows, fans). The general impact of the second type of adaptation is more limited because it strictly regards the single person (e.g. modify the clothing level). Finally, psychological adjustments are the less effective because they are related to the management of emotions, forcing an adaptation to the existing problems (e.g. ignore the problem) [12].

Among these strategies, only environmental modifications have a direct energy impact. However, some of them can quickly restore users' comfort but can be extremely inefficient in the energy perspective, dramatically affecting the energy consumptions. Researchers recorded many and frequent wasteful building operations that led to energy wastes by the heating, mechanical and electric system [13]. For example, Lindén et al. [4], reporting results from a survey of 600 households, discovered that only 17% of the subjects turn off lights when leaving a room and 40% air daily during wintertime.

However, in energy efficient buildings, users seem to pay more attention to their behaviours. In fact, researchers [11] found that occupants of such buildings are used to make less environmental and more personal adjustments to restore their comfort in comparison to their counterparts in standard buildings. It has been suggested that these greener actions could be a consequence of knowing to be in a certified and low-energy building [14].

2.3   Surveys on NZEB Performance

Since the expectations in relation to nZEBs' performances are extremely high both in the designers and the final users' perspectives, many researchers evaluated whether the design promises have been maintained.

Unfortunately, there is not a homogeneous response. In fact, some studies [10, 15] observed very good performance of low energy buildings since they have satisfactory indoor environments (e.g. ventilation requirements and temperature levels), use low energy and make users in a positive mood. On the contrary, other researchers [16, 17] underlined that to have an energy certification or to match the design requirements of nZEB are not always synonyms of good performance and excellent environment. Thermal conditions, in particular winter air temperature, have been advised as problematic issues [18]. Inadequate IAQ has been observed too [19]. In fact, the envelopes' tightening, drastically reducing the infiltrations, lead to a decrease in air quality whether the ventilation system does not correctly perform. Such problems are also a consequence of the lack of precise regulations to guarantee healthy environments.

Such divergences among the studies highlighted that the design evaluations often do not match the real ones, both in terms of IEQ and energy use. Researchers [10, 15, 20] that tried to understand the reasons that contribute to increasing the discrepancies between the design and the real assessments in nZEBs, suggested that the gap could be due to:

- Wrong assumptions during the design phase;
- Initial problems due to advanced energy-efficient technology;
- Peculiar users' interaction with building components;
- Uncertainty of building simulations in relation to input data and occupants' behaviours;
- High variability of users' behaviours due to psychological component.

All of the above-mentioned reasons provide to broaden the performance gap [17] but those connected to incorrect assumptions on users' personal and environmental adjustments have the greatest impact on the design, the simulation as well as the building operational phases. For this reasons, it is really important to estimate, in particular, the weight of the human component to improve energy and environmental performances. In recent years the international attention moved in this direction, promoting several projects concerning occupants' behaviours in the built environment [21, 22].

2.4   Simulations' Results

Simulations have been widely used to evaluate the building performances and also to estimate the impact of occupants' behaviours.

Discrepancies between real and predicted energy consumptions have been assessed along different building uses, locations and typologies [23, 24]. However, the identification of the energy performance gap in high performance buildings is even more relevant for the target these buildings should reach and for the clients' economic investments [25].

Large deviations between real and simulated results have been identified by several researchers. A mean discrepancy of about 20% has been assessed trough the comparison between actual and simulated energy uses of a representative stocks of LEED-certified buildings (i.e. 11 samples) [26]. Evaluations on a wider sample (i.e. 21 buildings) [27] highlighted a very small difference between actual and simulated performances among the buildings. However, a very high standard deviation (i.e. 46%) along the sample has been recognised: such variability is the sign of a big divergence in the performances of buildings with the same level of certification.

Further studies focused also on the effects of different users' attitudes on the building performance. It has been demonstrated that users' behaviours and management can significantly deteriorate the building performance and the energy generation [28]. Similarly, evaluations of the impact of different occupants' lifestyles in the nZEB context, showed that people are the key point of nZEBs' success [29]. In fact, if their management is wasteful, it is impossible to reach the nZEB target even if the building is of high performances.

References

1. Janda K (2011) Buildings don't use energy—people do! Planet earth 12–13. https://doi.org/10.3763/asre.2009.0050
2. Leth-Petersen S, Togeby M (2001) Demand for space heating in apartment blocks: measuring effects of policy measures aiming at reducing energy consumption. Energy Econ 23:387–403. https://doi.org/10.1016/S0140-9883(00)00078-5
3. Andersen RV, Toftum J, Andersen KK, Olesen BW (2009) Survey of occupant behaviour and control of indoor environment in Danish dwellings. Energy Build 41:11–16. https://doi.org/10.1016/j.enbuild.2008.07.004

4. Lindén AL, Carlsson-Kanyama A, Eriksson B (2006) Efficient and inefficient aspects of residential energy behaviour: what are the policy instruments for change? Energy Policy 34:1918–1927. https://doi.org/10.1016/j.enpol.2005.01.015
5. Brandemuehl MJ, Field KM (2011) Effects of variations of occupant behavior on residential building net zero energy performance. Proc Build Simul 14–16
6. Hong T, Yan D, D'Oca S, Chen C (2016) Ten questions concerning occupant behavior in buildings: the big picture. Build Environ 114:518–530. https://doi.org/10.1016/j.buildenv. 2016.12.006
7. Parliament of the European Union, Directive 2010/31/EU of 19 May 2010 on the energy performance of building (recast)
8. Parker D, Mills E, Rainer L et al (2012) Accuracy of the home energy saver energy calculation methodology. ACEEE Summer Study Energy Effic Build 1996:206–222
9. Poirazis H, Blomsterberg Å, Wall M (2008) Energy simulations for glazed office buildings in Sweden. Energy Build 40:1161–1170. https://doi.org/10.1016/j.enbuild.2007.10.011
10. Lenoir A, Cory S, Donn M, Garde F (2011) Users' behavior and energy performances of net zero energy buildings. Build Simul 14–16
11. Azizi NSM, Wilkinson S, Fassman E (2015) An analysis of occupants response to thermal discomfort in green and conventional buildings in New Zealand. Energy Build 104:191–198. https://doi.org/10.1016/j.enbuild.2015.07.012
12. Heerwagen J, Diamond RC (1992) Adaptations and coping: occupant response to discomfort in energy efficient buildings. Summer Study Energy Effic Build 1992:83–90
13. Heerwagen JH, Wise JA (1998) Green building benefits: differences in perceptions and experiences across manufacturing shifts. Heating, Pip air Cond 70:57–63
14. Khashe S, Heydarian A, Gerber D et al (2015) Influence of LEED branding on building occupants' pro-environmental behavior. Build Environ 94:477–488. https://doi.org/10.1016/j. buildenv.2015.10.005
15. Newsham G, Birt B, Arsenault C, et al (2012) Do green buildings outperform conventional buildings? Indoor environment and energy performance in North American offices
16. Sawyer L, De Wilde P, Turpin-Brooks S (2008) Energy performance and occupancy satisfaction: a comparison of two closely related buildings. Facilities 26:542–551. https://doi. org/10.1108/02632770810914299
17. Li C, Hong T, Yan D (2014) An insight into actual energy use and its drivers in high-performance buildings. Appl Energy 131:394–410. https://doi.org/10.1016/j.apenergy. 2014.06.032
18. Gou Z, Prasad D, Siu-Yu Lau S (2013) Are green buildings more satisfactory and comfortable? Habitat Int 39:156–161. https://doi.org/10.1016/j.habitatint.2012.12.007
19. Steinemann A, Wargocki P, Rismanchi B (2016) Ten questions concerning green buildings and indoor air quality. Build Environ:1–6. https://doi.org/10.1016/j.buildenv.2016.11.010
20. Santangelo A, Tondelli S (2017) Occupant behaviour and building renovation of the social housing stock: current and future challenges. Energy Build 145:276–283. https://doi.org/10. 1016/j.enbuild.2017.04.019
21. IEA EBC (2013) Annex 66: definition and simulation of occupant behavior in buildings
22. http://newtrend-project.eu/. (2017)
23. Menezes AC, Cripps A, Bouchlaghem D, Buswell R (2012) Predicted vs. actual energy performance of non-domestic buildings: using post-occupancy evaluation data to reduce the performance gap. Appl Energy 97:355–364. https://doi.org/10.1016/j.apenergy.2011.11.075
24. Majcen D, Itard LCM, Visscher H (2013) Theoretical versus actual energy consumption of labelled dwellings in the Netherlands: discrepancies and policy implications. Energy Policy 54:125–136. https://doi.org/10.1016/j.enpol.2012.11.008
25. Hong T, Taylor-Lange SC, D'Oca S et al (2016) Advances in research and applications of energy-related occupant behavior in buildings. Energy Build 116:694–702. https://doi.org/10. 1016/j.enbuild.2015.11.052
26. Turner C (2006) LEED Building performance in the Cascadia Region: a post occupancy evaluation report

27. Diamond R, Opitz M, Hicks T, et al (2006) Evaluating the energy performance of the first generation of LEED-certified commercial buildings
28. Bucking S, Athienitis A, Zmeureanu R (2011) Optimization of a Net-zero energy solar communities: effect of uncertainty due to occupant factors. ISES World Conf. Kassel, Germany, pp 1523–1528
29. Barthelmes VM, Becchio C, Corgnati SP (2016) Occupant behavior lifestyles in a residential nearly zero energy building: Effect on energy use and thermal comfort. Sci Technol Built Environ 22:960–975. https://doi.org/10.1080/23744731.2016.1197758

Chapter 3
Occupants' Behaviours Impact on Indoor Environment

Abstract People modify the environment by both their adaptive actions and their own presence. Perturbations due to human activities can be studied using decomposition methods since the environmental variables trends can be interpreted as temporal series. This approach has been applied to several parameters, recorded during an experimental survey, and the results are presented in this chapter. Statistical analyses have also been adopted to investigate users' influence according to different building uses.

3.1 Introduction

Occupants' presence and behaviours have an undeniable impact on the indoor environment [1]. In fact, when people remain in a room, even without interacting with the surrounding, they modify the indoor microclimate. The heat coming from the body and the breathing process affect especially the indoor temperature, humidity and CO_2 trends. In addition, when people actively interact with the building devices (e.g., opening and closing windows or switching the lights) their influence is even more considerable [2].

The following chapters present an investigation on occupants' influence through statistical analyses of environmental variables trend recorded in real cases study (Sect. 3.2), also trying to assess the impact in relation to different building typologies (Sect. 3.3).

3.2 Evaluation of Users' Perturbations

The environmental variables' trends can be interpreted as temporal series. In fact, time-series refer to data points taken over time that could have an internal structure (e.g., trend or seasonal variations) [3]. One of the most used techniques to analyse such data is the time series decomposition. This technique has been adopted in

© The Author(s) 2018

F. Stazi and F. Naspi, *Impact of Occupants' Behaviour on Zero-Energy Buildings*,
SpringerBriefs in Energy, https://doi.org/10.1007/978-3-319-71867-5_3

many fields and especially in the economic one [4, 5]. It consists in splitting the series into several independent components, each one representing a peculiar pattern. The four components are presented below:

- *Trend*: is the tendency of the considered phenomenon in relation to a specific period. This pattern exist when is present a long-term increase or decrease in the dataset;
- *Cycle*: is related to repeated fluctuations that are not related to fixed period;
- *Seasonal*: is composed of short term (e.g., daily or weekly) and regular movements that are cyclically repeated in the same period and, as a consequence, seasonality is always of a fixed and known period;
- *Random*: is also called residual effect since concerns sudden changes, which are unlikely to be repeated. It includes fluctuations that cannot be explained by trend and seasonal movements. In the perspective of assessing users' influence, this component concerns disturbances due to occupants' presence and actions.

The time series can be represented as the sum of these four components; however, since it is difficult to distinguish between the trend and the cycle components, they are usually combined [6]. Then, the time series Y_t can be expressed by the following equation:

$$Y_t = T_t + S_t + R_t \tag{3.1}$$

where Y_t, $t = 1,...,N$ represents the time series from which the components T_t, S_t and R_t have to be extracted. Figure 3.1 depicts the decomposition of indoor air

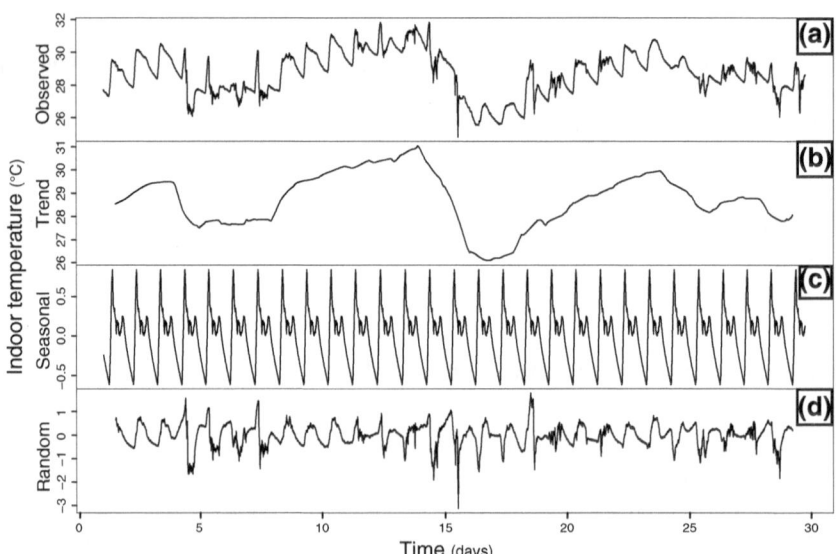

Fig. 3.1 Time series decomposition of indoor temperature (July 2016_Case study A)

temperature for one month (i.e., July 2016) for the case study A (Appendix A). The first trend (a) represents the observed one, while the followings (b, c and d) illustrate the trend, the seasonal and the random components, respectively.

The trend component follows the same tendency of the recorded temperature, varying between 26 and 31 °C. The seasonal shows a very little fluctuation (i.e., ± 0.5 °C), suggesting that, in the case study, such variations are minimal and can be omitted. The random component shows a greater variation: in general it is included in the range ± 1 °C but some peaks reached -3 °C. This tendency highlights the prominence of the random factor and so, the importance of its evaluation.

In this perspective, the random component of various environmental variables has been obtained decomposing observational data through the statistical software R studio [7]. In particular, to quantify occupants' impact, the random component has been analysed comparing samples related to occupied and empty periods.

An evaluation related to different climatic seasons is proposed in Fig. 3.2. It reports the analysis for four indoor parameters: (a) indoor temperature, (b) indoor humidity, (c) CO_2 concentration and (d) work-plane illuminance.

The boxplots related to occupied periods show a higher deviation from the median value (black solid lines) in comparison to empty ones in all the seasons. This highlights that occupants effectively influence the indoor environment. In relation to the season, the widest variations concern spring and autumn. In fact, during these periods the environmental conditions are extremely variable. Moreover, during such periods, the occupants frequently interact with the building devices to adjust the indoor environment [8]. On the contrary, on summer and wintertime occupants make fewer modifications to the surroundings (e.g., on summer the windows are usually kept open for all the working days) and the indoor environment is mainly affected by the outdoor one [9, 10].

The analysis demonstrates that users' impact is evident for all the variables. The dimension of the random component suggests that people can be considered more as a perturbation rather than a significant alteration of the variable trend. In fact, the random component variation on indoor temperature and CO_2 concentration is smaller than ± 1 °C and ± 100 ppm, respectively. Regarding the indoor humidity and illuminance, the oscillation is at most $\pm 6\%$ for the first and ± 150 lux for the second one. Such variations seem to indicate that in spaces intermittently occupied, the perturbation provoked by the users extinguish during unoccupied periods.

These conclusions suggest that the techniques adopted to predict the human component could be used without falling in big deviations from the real variables' trends.

3.3 Dependence on Building Usage Typology

The users' impact on the indoor environment can be also linked and analysed in relation to different building uses. The occupant density (persons/m^2) can be adopted as the parameter to discern the usage typologies [11]. In residential

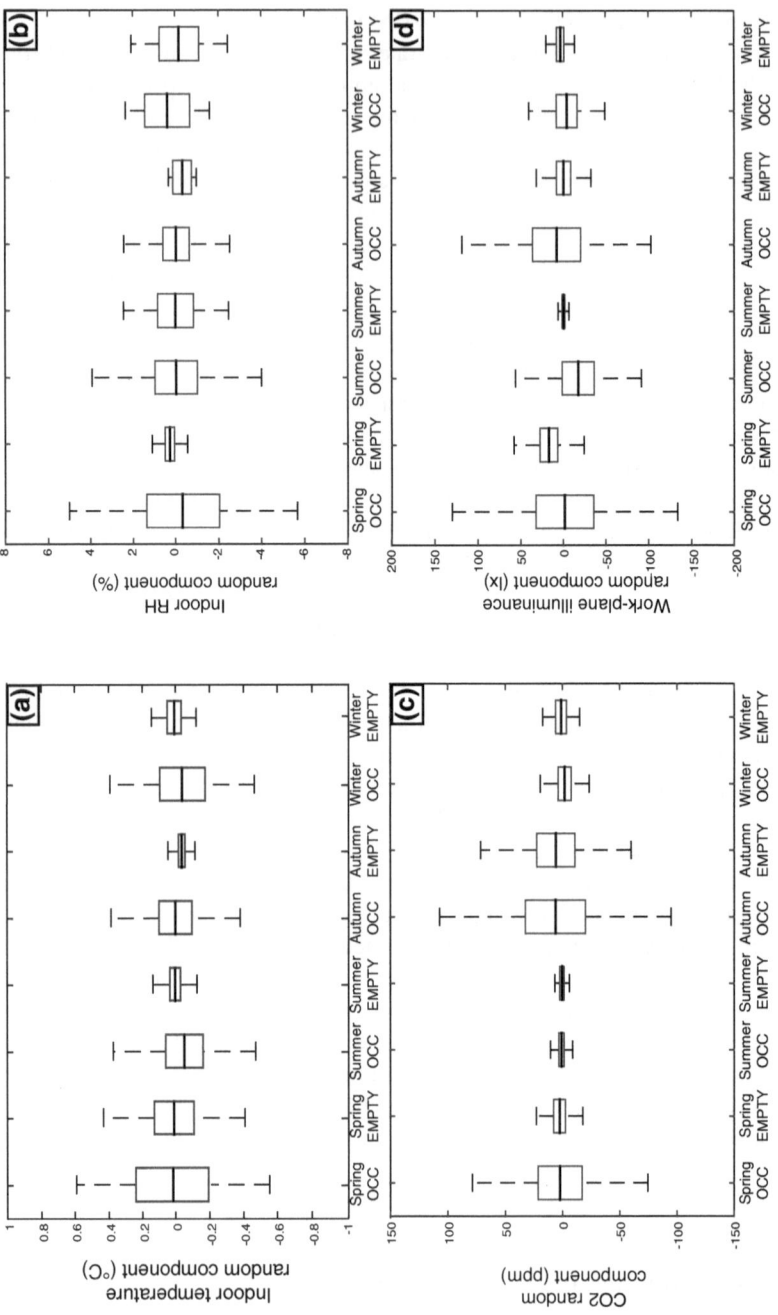

Fig. 3.2 Influence of the occupants' presence through the random component in relation to (**a**) indoor temperature (**b**) indoor humidity (**c**) CO_2 concentration (**d**) work plane illuminance

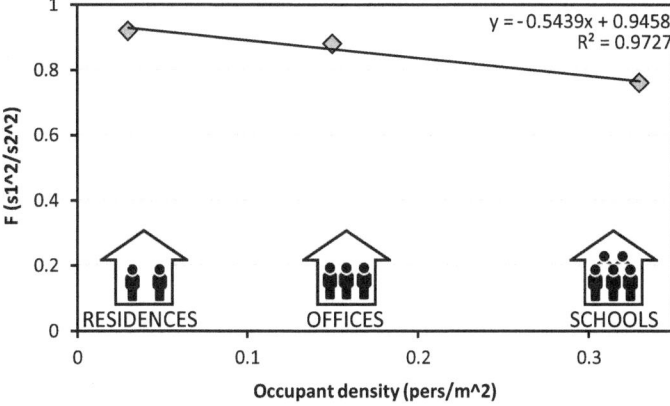

Fig. 3.3 F-test in relation to different occupant density

buildings, in fact, the density is usually low but the people have the highest level of interaction since they can freely adjust the systems and the devices. In offices, the density increases and the possibilities of adaptation decrease. Finally, in spaces with high density (e.g., school classrooms, conference rooms) occupants can take very few actions and they usually have to submit to building management.

To consider the users' impact on buildings with such different features, samples related to three specific cases study (refer to Appendix A) have been compared. The F-test between occupied and empty samples referred to indoor temperature has been performed for each of them. Finally, the F value is compared transversally.

The F-test is used to compare the variances of two samples (s_1^2 and s_2^2): the more the ratio is near to 1 the more the two samples belong to similar distributions [12]. Figure 3.3 shows that the lowest is the occupant density the nearest to 1 is the F-test. In residential buildings the F value is about 1 (F = 0.92), meaning that there are no substantial differences between occupied and unoccupied periods. For places with high occupant density, as classrooms, peoples' impact increases considerably (F = 0.76) but it should be noted that, in such contexts, peculiar decisional processes can play a key role (e.g., only the teacher decides the classroom management). Offices are placed in a mid-position: users' influence is evident (F = 0.88) and it should be predicted quite reliably.

References

1. Hong T, Yan D, D'Oca S, Chen C (2016) Ten questions concerning occupant behavior in buildings: the big picture. Build Environ 114:518–530. https://doi.org/10.1016/j.buildenv. 2016.12.006
2. Page J, Robinson D, Morel N, Scartezzini JL (2008) A generalised stochastic model for the simulation of occupant presence. Energy Build 40:83–98. https://doi.org/10.1016/j.enbuild. 2007.01.018

3. Mukhopadhyay SK (2015) Production planning and control: text and Cases, 3rd edn. PHI Learning Pvt. Ltd., Delhi
4. West M (1997) Time series decomposition. Biometrika 84:489–494. https://doi.org/10.1093/biomet/84.2.489
5. Beveridge S, Nelson CR (1981) A new approach to decomposition of economic time series into permanent and transitory components with particular attention to measurement of the "business cycle." J Monet Econ 7:151–174. https://doi.org/10.1016/0304-3932(81)90040-4
6. Bas MC, Ortiz J, Ballesteros L, Martorell S (2016) Analysis of the influence of solar activity and atmospheric factors on 7Be air concentration by seasonal-trend decomposition. Atmos Environ 145:147–157. https://doi.org/10.1016/j.atmosenv.2016.09.027
7. R Development Core Team (2011) R: A language and environment for statistical computing. Vienna, Austria : the R Foundation for Statistical Computing
8. Stazi F, Naspi F, D'Orazio M (2017) A literature review on driving factors and contextual events influencing occupants' behaviours in buildings. Build Environ 118:40–66. https://doi.org/10.1016/j.buildenv.2017.03.021
9. Herkel S, Knapp U, Pfafferott J (2008) Towards a model of user behaviour regarding the manual control of windows in office buildings. Build Environ 43:588–600. https://doi.org/10.1016/j.buildenv.2006.06.031
10. Zhang Y, Barrett P (2012) Factors influencing the occupants' window opening behaviour in a naturally ventilated office building. Build Environ 50:125–134. https://doi.org/10.1016/j.buildenv.2011.10.018
11. CEN Standard EN 15251 (2006) Indoor environmental input parameters for design and assessment of energy performance of buildings-addressing indoor air quality, thermal environment, lighting and acoustics Contents. CEN Brussels 1–52
12. Vakakis D (1984) Statistics: the F Test. TAFE Board

Chapter 4
Triggers for Users' Behaviours

Abstract People behaviours have an undeniable influence both on building performance and on indoor environmental quality. To predict their actions and obtain accurate predictions, it is of primary importance the understanding of their decision-making process. This chapter offers an overview of the main factors that stimulate users' behaviours. It firstly proposes an analysis regarding objective aspects (i.e., environmental variables, time-related events and contextual factors), aiming at underlying the triggers for each adaptive action. Then, it focuses on subjective aspects (i.e., physiological, psychological and social factors), in particular, highlighting how people's attitudes change in the nZEB contexts.

4.1 Introduction

One of the major unresolved issues regarding the energy-related behaviours concerns the interaction between occupants and building devices (e.g., windows, lights). In fact, since comfort conditions are individually defined, people modify the environment according to personal preferences and needs. This means that users' behaviours are not driven by deterministic rules but their adjustments vary according to both individual and contextual stimuli.

Many studies [1–3] have been carried out with the aim to understand the factors that trigger users' actions and to classify them. The different aspects can be sub-divided in two macro-categories: the first concerns concrete and objective characteristics which are linked to measurable aspects, while the second is related to personal and individual features which are peculiar for each person. The former macro-category includes environmental, time-related and contextual factors. The latter concerns physiological, psychological and social features. In addition to these classes, Peng et al. [1] included also the "random" category to consider unpredictable actions which depend on uncertain factors or follow unknown rules.

Figure 4.1 displays the above-mentioned triggers, subdivided between the two macro categories, and reports several examples for each aspect.

© The Author(s) 2018
F. Stazi and F. Naspi, *Impact of Occupants' Behaviour on Zero-Energy Buildings*,
SpringerBriefs in Energy, https://doi.org/10.1007/978-3-319-71867-5_4

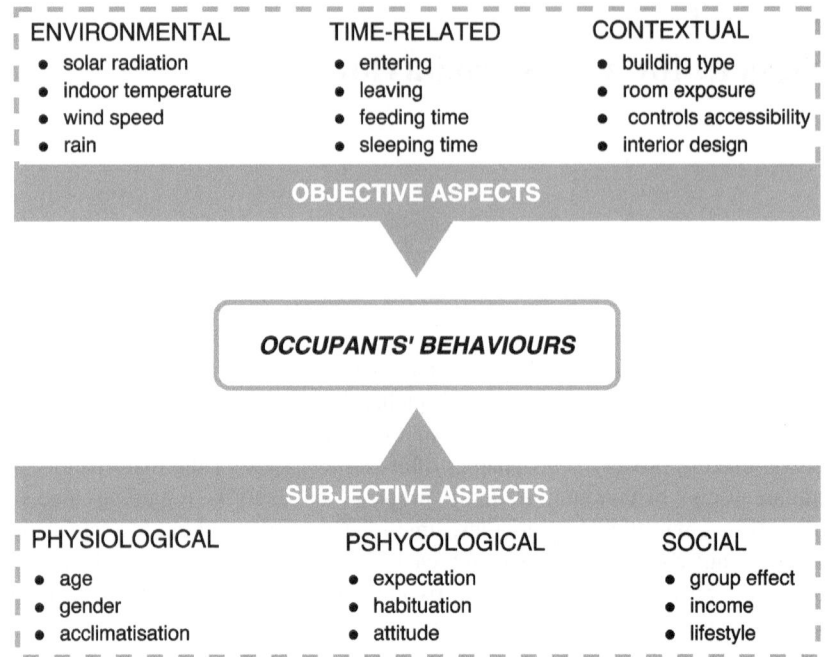

Fig. 4.1 Categorization of the triggers for users' adaptation

The following sub-paragraphs examine the above-mentioned drivers. Section 4.2 focuses on objective aspects, while Sect. 4.3 analyses personal features exploring, in particular, how users' psychology modifies in nZEBs.

4.2 Objective Factors

The stimuli related to environmental, time-related and contextual factors have been widely analysed for their undoubted affection on occupants' actions and for the easiness in recording and comparing them. Among them, environmental and time-related factors directly and instantaneously influence people's perceptions [4]. In additions, such variables have a great reliability since they are related only to acquisition systems' accuracy.

Environmental Variables
Environmental variables are the main triggers for almost all the users' adaptive actions. Some parameters are strictly related to thermal or visual comfort, while others are mainly connected to IAQ requirements.

Indoor and outdoor temperatures have been recognised as two of the most important stimuli connected to *thermal comfort*. For this reason, they drive several adaptive actions: window [5, 6], air-conditioning [7], thermostat [8] and fan use [9, 10]. Few studies [9, 11] found a connection also with shading adjustments. While, only the indoor temperature has been recognised as a stimulus for the doors use [10, 12].

Whether indoor or outdoor parameters are the most influencing ones is still an open question [4]. Some authors affirm that outdoor variables should be preferred since they are independent of users' adjustments. On the contrary, other researchers assert that indoor parameters have to be chosen since the perceptions of people who live inside buildings should be mainly influenced by the indoor environment [10].

Despite such divergences, in general, when both indoor and outdoor temperatures rise, the probability of opening windows [13, 14], of switch on the air-conditionings [15] and the fans and to lower the blinds [12] increase as well. Conversely, their reduction leads to higher frequencies in window closing [16] and in turning the heating on [17]. Figure 4.2a shows how the outdoor temperature modify the window opening probability in offices. It can be noted that the probability becomes relevant when the variable overcomes 10 °C and it is almost certain for temperatures higher than 28 °C.

The surveys promoted to assess the triggers related to *visual comfort* did not bring to definite results jet. In fact, illuminance levels [18, 19], solar radiation [20, 21] and glare [11, 22] have been recognised as influencing factors for blinds and shadings adjustments. However, the distinct physical meaning of such parameters makes comparisons not so immediate. Similarly, illuminance [19, 23] has been identified as the main trigger for light switching behaviours, but which type of illuminance (e.g., global, work-plane) has to be preferred is still an unresolved issue.

Regardless the differences among the studies and the proposed results, researchers agree in affirming that the decrease of illuminance levels directly trigger switch-on actions [24], while their increase influence blinds opening [25]. Similar conclusions have been made also in relation to solar radiation and glare [26].

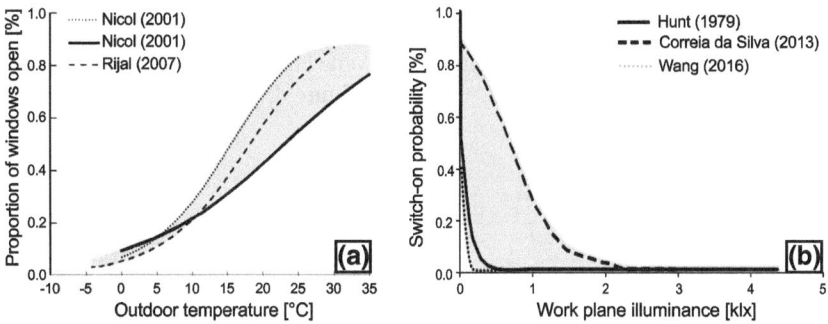

Fig. 4.2 Influence of environmental variables on interaction with devices in offices: (**a**) window opening triggered by outdoor temperature and (**b**) light-switching driven by work plane illuminance

Figure 4.2b depicts the increasing of light-switching probability according to work plane illuminance decrease in offices.

Indoor air quality and ventilation are usually evaluated using CO_2 concentration [27]. This parameter has been studied in several building uses but it has been recognised as a trigger for window opening only in the residential context [28, 29]. The CO_2 growth often leads to a window opening increase [16]; however many studies [30–32] also reported that users are not aware of such increase, and, as a consequence, they are used to suffer high CO_2 levels [33].

In addition to indoor and outdoor temperature and CO_2 concentration, *other variables* have been cited as relevant stimuli. Rainfall, wind speed and direction can drive windows adjustments [34, 35], even if they are strictly connected to building shape and window typology. Few studies reported also the indoor and outdoor relative humidity among the drivers for windows usage [29, 35], AC use [36] and thermostat control [36, 37].

Time-Related Factors

Time-related factors have been deeply analysed too. This category concerns events that are connected to the specific time of the day: for example, the arrival at work or the sleeping time at home. In *offices*, arrival and departure periods are the most influencing [35]. At arrival, it has been recorded the highest frequency of window opening [38], lights up [39] and shading adjustments [20]. Conversely, departures are characterised especially by window closing [40] and turning off lights [20]. Breaks and arrivals also drive students in *school classrooms*, in fact, several studies recorded an increase of such actions during these periods [41]. Intermediate periods, in both these building uses, are marked by very little interactions with building devices since users tend to adapt to indoor conditions and so, they are less inclined to intervene [31, 42].

In *residential buildings*, specific periods are connected to some peculiar actions. During mornings and evenings have been recorded peaks in window state changes since they are related to specific activities, as cooking [14, 29]. Sleeping and eating times trigger also the AC modifications [7, 43], increasing the switch-on proba-bility. Turn-off behaviours are more frequent after getting up and leaving the room. A different use pattern has been noted also between weekdays and weekends [44]: the first have peaks after 6 p.m., when people come back home, while in the second, adjustments are made along all the day. Figure 4.3 highlights how change

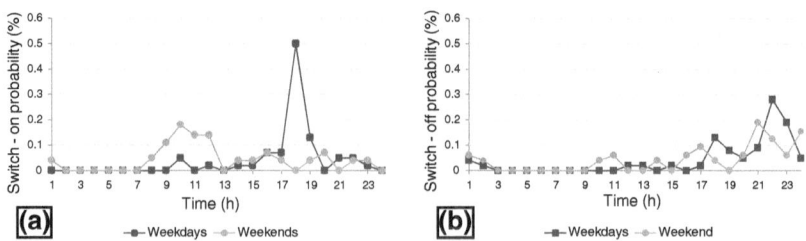

Fig. 4.3 Influence of time of the day and day of the week on the air conditioning use: (**a**) switch-on probability; (**b**) switch-off probability

occupants' interaction with AC units in relation to time of the days and during weekdays/weekends.

It is evident that users are influenced by both environmental stimuli and daily/ weekly routine. Hence, the identification of occupants' habits is a key point in the behavioural assessment.

Contextual Factors
Contextual factors have been less studied than environmental and time-related ones but their importance in users' behaviours has been identified by several studies [45, 46]. In particular, a comprehensive review, proposed by O'Brien and Gunay [47], highlights that availability and accessibility of adaptive measures, interior design, mechanical/electrical systems and outdoor views actively contribute to the occupants' decision-making process. Since very few studies investigate such aspects and, when it happens, information are usually fragmentary and incomplete, the authors also proposed a framework to systematic report such observations. In fact, a standardised representation of contextual factors will allow their inclusion in behavioural models and in BEPS tools.

4.3   Personal Features

Subjective aspects concern features and perceptions connected to the people's personal sphere. Surveys, questionnaires and individual investigations are essential methods to acquire this kind of data. However, this massive intrusion in subjects' life can make users reluctant to provide personal information and can also cause privacy issues [48].

Despite these obstacles, many researchers investigated subjective aspects both in nZEB and in "standard" buildings, also aiming at assessing how users' perceptions change in these two contexts.

Physiological Factors
As anticipated in Fig. 4.1, physiological adaptations concern the strategies the human body use to acclimatise to discomfort [2]. In this perspective, *age* and *gender* are the two main aspects that influence people's physical perception. Since the ageing compromises body defences from cold [49], it was noted that older subjects (e.g., age >60) usually prefer a warmer environment in comparison to younger people [50, 51]. Studies conducted in schools [30, 52] found that children are more sensitive to high temperature than adults, leading to overestimations in comfort temperature. However, it has been suggested that such difference could be a consequence of further aspects, typical of the scholastic environment (e.g., practical activities, overcrowding) [53].

Researchers which analysed gender thermal preferences seem to agree in assessing that women prefer warmer temperatures compared to men [54]. This predilection towards warm could be also a consequence of women morphological aspects that influence heat balance, thermoregulation and thermal perceptions [51].

Psychological Factors

Along with physical features, also psychological aspects are extremely important in people's behaviours. Such type of adaptation is influenced by past experiences and expectations, which can be also conditioned from cultural and cognitive aspects [55]. The psychological component can largely affect thermal comfort, inclinations and energy-related behaviours [56].

Studies carried out in low-energy buildings found that, in general, people who live and work in such environment are more comfortable and satisfied than users of standard buildings [57]. Even if indoor conditions between these two typologies are very similar, sometimes a "green" label seems to have the ability to improve users' perceptions of the indoor environment [58]. However, such conclusion cannot be generally applied since some authors reported that users in LEED-certified buildings are not more satisfied with the environment than users in standard ones [59].

Despite some divergences among the studies, occupants of green buildings appear to be more tolerant in relation to IEQ, adapting to great discomfort levels [57, 60, 61]. In fact, in buildings where the interaction with the devices is freely allowed, users tend to expand their comfort zone [62]. Such psychological modification, called "*Forgiveness factor*" [63], seems to be higher for occupants of green buildings that understand how the building and the controls work [64].

Such studies suggest that perceptions of people in high-performance buildings undergo some modifications. However, these alterations can cause opposite consequences on the final energy use. On one side, "positive perceptions" make users more energy conscious in their behaviours. In fact, they tend to adopt green solutions to enhance the building performance, even threatening their own comfort [65]. This kind of actions is called *pro environmental behaviours* and concern behaviours that harm the environment as little as possible (e.g., the choice of natural light to electric one) [66]. On the other side, "negative perceptions" are characterised by the *rebound effect* (also called Jevons' paradox [67]), which effects have been divided into direct and indirect. Direct effects are a consequence of the improvement of the efficiency of an energy service: such enhancement causes a higher demand than before because the service has become cheaper for the users. Indirect effects start from the premise that an energy service has been improved both in the energy and economic perspectives and so, the users can utilise the saved money for a new service that requires energy, though [68].

In general, the improvement of occupants' behaviours can be promoted both at personal and building level. The first concern a direct influence on the user psychological and emotional spheres (e.g., education, information, social influences). The second regards structural innovations that modify then surrounding (e.g., new appliance, structures) [69].

Since it has been widely assessed that architectural solutions alone are insufficient to achieve energy target, several campaigns have been promoted to improve occupants' energy behaviours [70, 71]. Moreover, studies that investigated strategies to directly influence behaviours found that users that have been frequently educated and trained in the correct use of the controls maximise both the building energy efficiency and their own satisfaction [65]. However, whether information and training are sporadic, very few improvements have been reached. In fact, the users' motivation and participation quickly decay if continuous reinforcements are absent [72].

Social Factors

People are also influenced by *social* forces. Working and living with other subjects, a user can modify its personal habits according to the building or group rules, so it is extremely important encouraging and spreading employees towards positive behaviours [65].

It is evident that subjective aspects are important as well as objective ones. However, a difficulty in recording and quantifying them is still undeniable. These limitations underline the evident necessity for a *multidisciplinary approach*. In fact, to reach a comprehensive understanding of behavioural adaptations and to transpose them in mathematical models, it would be advantageous the collaboration of different research fields, as engineering, social and psychological sciences. In this direction, some studies [73, 74] aimed at applying models and theories from psychology and sociology to conceptualise the human interactions with buildings. Therefore, the collaboration of reliable measurements and social science is an essential step for the recording of non-objective variables, the realisation of suitable questionnaires and so, to improve the entire survey methodology [48].

References

1. Peng C, Yan D, Wu R, et al (2012) Quantitative description and simulation of human behavior in residential buildings. Build Simul 5:85–94. https://doi.org/10.1007/s12273-011-0049-0
2. Inkarojrit V (2005) Balancing comfort: occupants' control of window blinds in private offices. University of California, Berkeley
3. Fabi V, Andersen RV, Corgnati S, Olesen BW (2012) Occupants' window opening behaviour: a literature review of factors influencing occupant behaviour and models. Build Environ 58:188–198. https://doi.org/10.1016/j.buildenv.2012.07.009
4. Stazi F, Naspi F, D'Orazio M (2017) A literature review on driving factors and contextual events influencing occupants' behaviours in buildings. Build Environ 118:40–66. https://doi.org/10.1016/j.buildenv.2017.03.021
5. D'Oca S, Hong T (2014) A data-mining approach to discover patterns of window opening and closing behavior in offices. Build Environ 82:726–739. https://doi.org/10.1016/j.buildenv.2014.10.021
6. Schweiker M, Haldi F, Shukuya M, Robinson D (2012) Verification of stochastic models of window opening behaviour for residential buildings. J Build Perform Simul 5:55–74. https://doi.org/10.1080/19401493.2011.567422
7. Habara H, Yasue R, Shimoda Y (2013) Survey on the occupant behavior relating to window and air conditioner operation in the residential buildings. In: 13th conference of the international building performance simulation association, pp 2007–2013
8. Lin B, Wang Z, Liu Y, et al (2016) Investigation of winter indoor thermal environment and heating demand of urban residential buildings in China's hot summer—cold winter climate region. Build Environ 101:9–18. https://doi.org/10.1016/j.buildenv.2016.02.022

9. Raja IA, Nicol JF, McCartney KJ, Humphreys MA (2001) Thermal comfort: use of controls in naturally ventilated buildings. Energy Build 33:235–244. https://doi.org/10.1016/S0378-7788(00)00087-6
10. Rijal HB, Tuohy PG, Nicol JF et al (2008) Development of adaptive algorithms for the operation of windows, fans and doors to predict thermal comfort and energy use in Pakistani buildings. ASHRAE Trans 114:555–573
11. Sutter Y, Dumortier D, Fontoynont M (2006) The use of shading systems in VDU task offices: a pilot study. Energy Build 38:780–789. https://doi.org/10.1016/j.enbuild.2006.03.010
12. Haldi F, Robinson D (2008) On the behaviour and adaptation of office occupants. Build Environ 43:2163–2177. https://doi.org/10.1016/j.buildenv.2008.01.003
13. Li N, Li J, Fan R, Jia H (2015) Probability of occupant operation of windows during transition seasons in office buildings. Renew Energy 73:84–91. https://doi.org/10.1016/j.renene.2014.05.065
14. Jeong B, Jeong J-W, Park JS (2016) Occupant behavior regarding the manual control of windows in residential buildings. Energy Build 127:206–216. https://doi.org/10.1016/j.enbuild.2016.05.097
15. Nicol JF, Humphreys MA (2004) A stochastic approach to thermal comfort—occupant behavior and energy use in buildings. ASHRAE Trans 110:Part II: 68–554
16. Fabi V, Andersen RV, Corgnati SP (2012) Window opening behaviour : simulations of occupant behaviour in residential buildings using models based on a field survey. In: 7th Windsor conference: the changing context of comfort an unpredictable world, pp 12–15
17. Schweiker M, Shukuya M (2009) Comparison of theoretical and statistical models of air-conditioning-unit usage behaviour in a residential setting under Japanese climatic conditions. Build Environ 44:2137–2149. https://doi.org/10.1016/j.buildenv.2009.03.004
18. Vine E, Lee E, Clear R, et al (1998) Office worker response to an automated Venetian blind and electric lighting system: a pilot study. Energy Build 28:205–218. https://doi.org/10.1016/S0378-7788(98)00023-1
19. Nicol F, Wilson M, Chiancarella C (2006) Using field measurements of desktop illuminance in European offices to investigate its dependence on outdoor conditions and its effect on occupant satisfaction, and the use of lights and blinds. Energy Build 38:802–813. https://doi.org/10.1016/j.enbuild.2006.03.014
20. Correia da Silva P, Leal V, Andersen M (2013) Occupants interaction with electric lighting and shading systems in real single-occupied offices: results from a monitoring campaign. Build Environ 64:152–168. https://doi.org/10.1016/j.buildenv.2013.03.015
21. Yao J (2014) Determining the energy performance of manually controlled solar shades: A stochastic model based co-simulation analysis. Appl Energy 127:64–80. https://doi.org/10.1016/j.apenergy.2014.04.046
22. Maniccia D, Rutledge B, Rea M, Morow W (1999) Occupant use of manual lighting controls in private offices. Iesna 489–512. https://doi.org/10.1080/00994480.1999.10748274
23. Hunt DRG (1979) The use of artificial lighting in relation to daylight levels and occupancy. Build Environ 14:21–33. https://doi.org/10.1016/0360-1323(79)90025-8
24. Reinhart CF, Voss K (2003) Monitoring manual control of electric lighting and blinds. Light Res Technol 35:243–258. https://doi.org/10.1191/1365782803li064oa
25. Haldi F, Robinson D (2009) A comprehensive stochastic model of blind usage: theory and validation. Build. Simul. Conf. Glasgow, Scotland, pp 545–552
26. Gunay HB, O'Brien W, Beausoleil-Morrison I, Gilani S (2016) Development and implementation of an adaptive lighting and blinds control algorithm. Build Environ 113: http://doi.org/10.1016/j.buildenv.2016.08.027
27. Salthammer T, Uhde E, Schripp T, et al (2016) Children's well-being at schools: impact of climatic conditions and air pollution. Environ Int 94:196–210. https://doi.org/10.1016/j.envint.2016.05.009

28. Andersen R, Fabi V, Toftum J, et al (2013) Window opening behaviour modelled from measurements in Danish dwellings. Build Environ 69:101–113. https://doi.org/10.1016/j.buildenv.2013.07.005

29. Calì D, Andersen RK, Müller D, Olesen B (2016) Analysis of occupants' behavior related to the use of windows in German households. Build Environ 103:54–69. https://doi.org/10.1016/j.buildenv.2016.03.024

30. Stazi F, Naspi F, Ulpiani G, Perna C Di (2017) Indoor air quality and thermal comfort optimization in classrooms developing an automatic system for windows opening and closing. Energy Build 139:732–746. https://doi.org/10.1016/j.enbuild.2017.01.017

31. Santamouris M, Synnefa A, Asssimakopoulos M et al (2008) Experimental investigation of the air flow and indoor carbon dioxide concentration in classrooms with intermittent natural ventilation. Energy Build 40:1833–1843. https://doi.org/10.1016/j.enbuild.2008.04.002

32. Dias Pereira L, Raimondo D, Corgnati SP, Gameiro da Silva M (2014) Assessment of indoor air quality and thermal comfort in Portuguese secondary classrooms: Methodology and results. Build Environ 81:69–80. https://doi.org/10.1016/j.buildenv.2014.06.008

33. Daisey JM, Angell WJ, Apte MG (2003) Indoor air quality, ventilation and health symptoms in schools: an analysis of existing information. Indoor Air 13:53–64. https://doi.org/10.1034/j.1600-0668.2003.00153.x

34. Johnson T, Long T (2005) Determining the frequency of open windows in residences: a pilot study in Durham, North Carolina during varying temperature conditions. J Expo Anal Environ Epidemiol 15:329–349. https://doi.org/10.1038/sj.jea.7500409

35. Zhang Y, Barrett P (2012) Factors influencing the occupants' window opening behaviour in a naturally ventilated office building. Build Environ 50:125–134. https://doi.org/10.1016/j.buildenv.2011.10.018

36. Bae C, Chun C (2009) Research on seasonal indoor thermal environment and residents' control behavior of cooling and heating systems in Korea. Build Environ 44:2300–2307. https://doi.org/10.1016/j.buildenv.2009.04.003

37. Andersen R, Olesen B, Toftum J (2011) Modelling occupants' heating set-point prefferences. In: 12th Conference of the international building performance simulation association, Sydney, 14–16 November, pp 151–156

38. Herkel S, Knapp U, Pfafferott J (2008) Towards a model of user behaviour regarding the manual control of windows in office buildings. Build Environ 43:588–600. https://doi.org/10.1016/j.buildenv.2006.06.031

39. Boyce PR, Veitch JA, Newsham GR et al (2006) Occupant use of switching and dimming controls in offices. Light Res Technol 38:358–376. https://doi.org/10.1177/1477153506070994

40. Yun GY, Steemers K (2008) Time-dependent occupant behaviour models of window control in summer. Build Environ 43:1471–1482. https://doi.org/10.1016/j.buildenv.2007.08.001

41. De Giuli V, Da Pos O, De Carli M (2012) Indoor environmental quality and pupil perception in Italian primary schools. Build Environ 56:335–345. https://doi.org/10.1016/j.buildenv.2012.03.024

42. Fritsch R, Kohler a., Nygård-Ferguson M, Scartezzini J-L (1990) A stochastic model of user behaviour regarding ventilation. Build Environ 25:173–181. https://doi.org/10.1016/0360-1323(90)90030-U

43. Ren X, Yan D, Wang C (2014) Air-conditioning usage conditional probability model for residential buildings. Build Environ 81:172–182. http://doi.org/10.1016/j.buildenv.2014.06.022

44. Kempton W, Feuermann D, McGarity AE (1992) "I always turn it on super": user decisions about when and how to operate room air conditioners. Energy Build 18:177–191. https://doi.org/10.1016/0378-7788(92)90012-6

45. Von Grabe J (2016) The systematic identification and organization of the context of energy-relevant human interaction with buildings—a pilot study in Germany. Energy Res Soc Sci 12:75–95. https://doi.org/10.1016/j.erss.2015.12.001

46. Jones RV, Fuertes A, Gregori E, Giretti A (2017) Stochastic behavioural models of occupants' main bedroom window operation for UK residential buildings. Build Environ 118:144–158. https://doi.org/10.1016/j.buildenv.2017.03.033

47. O'Brien W, Gunay HB (2014) The contextual factors contributing to occupants' adaptive comfort behaviors in offices—a review and proposed modeling framework. Build Environ 77:77–88. https://doi.org/10.1016/j.buildenv.2014.03.024

48. Hong T, Yan D, D'Oca S, Chen C (2016) Ten questions concerning occupant behavior in buildings: the big picture. Build Environ 114:518–530. https://doi.org/10.1016/j.buildenv. 2016.12.006

49. Kenney WL, Munce TA (2003) Physiology of aging invited review: aging and human temperature regulation. J Appl Physiol 95:2598–2603

50. Indraganti M, Rao KD (2010) Effect of age, gender, economic group and tenure on thermal comfort: A field study in residential buildings in hot and dry climate with seasonal variations. Energy Build 42:273–281. https://doi.org/10.1016/j.enbuild.2009.09.003

51. Mishra AK, Ramgopal M (2013) Field studies on human thermal comfort—an overview. Build Environ 64:94–106. https://doi.org/10.1016/j.buildenv.2013.02.015

52. Teli D, Jentsch MF, James PAB (2012) Naturally ventilated classrooms: an assessment of existing comfort models for predicting the thermal sensation and preference of primary school children. Energy Build 53:166–182. https://doi.org/10.1016/j.enbuild.2012.06.022

53. Dear R De, Kim J, Candido C, Deuble M (2014) Summertime thermal comfort in Australian School Classrooms. In: 8th Windsor Conference, p 21

54. Karjalainen S (2007) Gender differences in thermal comfort and use of thermostats in everyday thermal environments. Build Environ 42:1594–1603. https://doi.org/10.1016/j. buildenv.2006.01.009

55. Brager G, Dear R de (1998) Thermal adaptation in the built environment: a literature review. Energy Build. 17:83–96

56. Sovacool BK (2009) Rejecting renewables: the socio-technical impediments to renewable electricity in the United States. Energy Policy 37:4500–4513. https://doi.org/10.1016/j.enpol. 2009.05.073

57. Gou Z, Prasad D, Siu-Yu Lau S (2013) Are green buildings more satisfactory and comfortable? Habitat Int 39:156–161. https://doi.org/10.1016/j.habitatint.2012.12.007

58. Holmgren M, Kabanshi A, Sörqvist P (2017) Occupant perception of "green" buildings: distinguishing physical and psychological factors. Build Environ 114:140–147. https://doi. org/10.1016/j.buildenv.2016.12.017

59. Altomonte S, Schiavon S (2013) Occupant satisfaction in LEED and non-LEED certified buildings. Build Environ 68:66–76. https://doi.org/10.1016/j.buildenv.2013.06.008

60. Steinemann A, Wargocki P, Rismanchi B (2017) Ten questions concerning green buildings and indoor air quality. Build Environ 112:351–358. https://doi.org/10.1016/j.buildenv.2016.11.010

61. Azizi NSM, Wilkinson S, Fassman E (2015) An analysis of occupants response to thermal discomfort in green and conventional buildings in New Zealand. Energy Build 104:191–198. https://doi.org/10.1016/j.enbuild.2015.07.012

62. Nicol JF, Humphreys MA (2002) Adaptive thermal comfort and sustainable thermal standards for buildings. Energy Build 34:563–572. https://doi.org/10.1016/S0378-7788(02)00006-3

63. Humphreys MA (2005) Quantifying occupant comfort: are combined indices of the indoor environment practicable? Build Res Inf 33:317–325. https://doi.org/10.1080/ 09613210500161950

64. Deuble MP, de Dear RJ (2012) Green occupants for green buildings: the missing link? Build Environ 56:21–27. https://doi.org/10.1016/j.buildenv.2012.02.029

65. Day JK, Gunderson DE (2015) Understanding high performance buildings: the link between occupant knowledge of passive design systems, corresponding behaviors, occupant comfort and environmental satisfaction. Build Environ 84:114–124. https://doi.org/10.1016/j. buildenv.2014.11.003

66. Steg L, Vlek C (2009) Encouraging pro-environmental behaviour: an integrative review and research agenda. J Environ Psychol 29:309–317. https://doi.org/10.1016/j.jenvp.2008.10.004

67. John MP, Kozo M, Mario G, Blake A (2008) The Jevons Paradox and the myth of resource efficiency improvements. London
68. Calì D, Osterhage T, Streblow R, Müller D (2016) Energy performance gap in refurbished German dwellings: lesson learned from a field test. Energy Build 127:1146–1158. https://doi.org/10.1016/j.enbuild.2016.05.020
69. Santangelo A, Tondelli S (2017) Occupant behaviour and building renovation of the social housing stock: current and future challenges. Energy Build 145:276–283. http://doi.org/10.1016/j.enbuild.2017.04.019
70. Pisello AL, Asdrubali F (2014) Human-based energy retrofits in residential buildings: a cost-effective alternative to traditional physical strategies. Appl Energy 133:224–235. https://doi.org/10.1016/j.apenergy.2014.07.049
71. Fabi V, Barthelmes VM, Corgnati SP (2016) Impact of an engagement campaign on user behaviour change in office environment. In: Indoor Air 2016—14th International conference on indoor air quality and climate. Ghent, Belgium, pp 1–8
72. Moloney S, Horne RE, Fien J (2010) Transitioning to low carbon communities-from behaviour change to systemic change: lessons from Australia. Energy Policy 38:7614–7623. https://doi.org/10.1016/j.enpol.2009.06.058
73. von Grabe J (2016a) Decision models and data in human-building interactions. Energy Res Soc Sci 19:61–65. https://doi.org/10.1016/j.erss.2016.05.022
74. von Grabe J (2016b) How do occupants decide their interactions with the building? From qualitative data to a psychological framework of human-building-interaction. Energy Res Soc Sci 14:46–60. https://doi.org/10.1016/j.erss.2016.01.002

Chapter 5
Occupants' Adaptive Actions

Abstract People adapt the indoor environment to restore their preferred comfort sensations intervening on building's controls. Such adjustments have a great impact on the building performances, especially in the nZEB contexts. As a consequence, the investigation and understanding of human-building interaction patterns can help to bridge the gap between real and predicted energy consumptions and to design buildings tailored for users' preferences and needs. In this perspective, the current chapter investigates many adaptive actions, highlighting, in particular, the most recurring and frequent behaviours the occupants perform in buildings.

5.1 Introduction

Building occupants can usually interact with many different devices to improve the indoor environmental quality and their actions have a large effect on the overall building performance. Moreover, their behaviours change according to the specific device because of the different device's features, environmental drivers and habits. Figure 5.1 sketches the cycle of interactions between the user and the environment, focusing on the action he can take to restore the preferred comfort conditions and satisfy the needs.

This chapter explores the interactions between the users and seven typical building devices. Each sub-chapter, at first, reports the literature which focused on the topic, highlighting the surveyed geographic areas, building uses, monitoring periods and durations. Then, it illustrates how much the actions can influence buildings' energy consumptions and users' comfort sensations. It analyses the main factors that trigger occupants' behaviours and attitudes, also reporting results from real cases study. Finally, it briefly presents the behavioural models developed to predict that specific interaction (more precise evaluations on the behavioural models' development are presented in Chap. 7). The final sub-chapter focuses on the identification of the sequence of behaviours when the users can take different actions to reach the same goal.

© The Author(s) 2018   31
F. Stazi and F. Naspi, *Impact of Occupants' Behaviour on Zero-Energy Buildings*,
SpringerBriefs in Energy, https://doi.org/10.1007/978-3-319-71867-5_5

Fig. 5.1 Cycle of interaction between the user and the environment: external and internal triggers, in association to peculiar user's comfort requirements affect the user perception. As a consequence, occupants interact with building devices to satisfy their needs. Such modification alter the indoor environment and the also the user perception

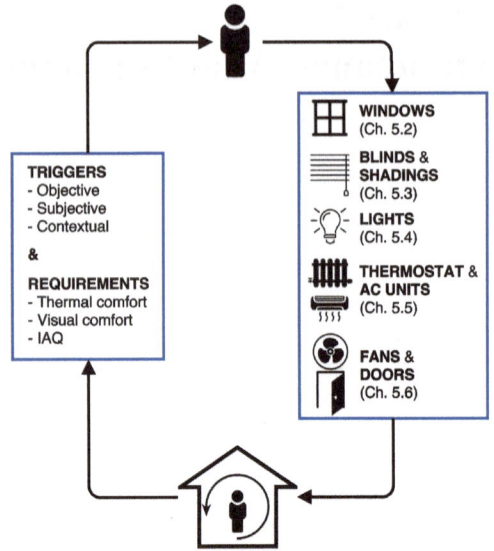

5.2   Window Opening and Closing

Window use has been one of the most studied patterns since the literature on this topic covers more than 25 years. The surveys have been carried out in many countries (e.g., USA, Europe, China), climates (e.g., Mediterranean, continental, hot-humid) and building uses (i.e., residential, offices and schools) and through different monitoring periods (e.g., only a specific thermal season) and durations (e.g., few months, several years). Human-windows interaction attracted the researchers' interest for a main reason: operations on windows considerably impact on both building energy consumptions and users' thermal comfort [1, 2].

Energy Consumptions

Energy consumptions can be influenced by people's adjustments during both the heating and non-heating season. In fact, even short opening actions can produce a sensible perturbation on the indoor temperature, especially when the indoor-outdoor temperature difference is large. Studies found that proper ventilation measures can save HVAC energy up to 47% [3] or increase thermal energy up to 66% [4]. In this perspective, improving occupants' behaviours is an essential step to promote building energy efficiency and reduce energy waste [5].

Comfort Perspective

Careful windows control can also provide optimisations of users' thermal comfort. Window openings and closings have the duty to suddenly modify the indoor environment both in terms of temperature and IAQ. As a consequence, people are motivated to adjust the window status. In addition, when occupants can freely adapt windows position, they tend to accept a wider temperatures range than people forced to undergo the thermal environment [6].

Triggering Parameters

Due to the wide literature concerning this topic, a great number of triggering parameters has been identified. However, the differences in monitoring and analyses make precise results still missing. Despite this lack, some factors have been clearly identified as influencing. Among the *environmental parameters*, indoor and outdoor temperature and CO_2 concentration have been assessed as the main drivers. Figure 5.2 shows the envelopes of several behavioural models for window opening from a literature review [7]. It can be noted that the envelopes cover a wide range of temperatures but over 10 °C for the outdoor and 15 °C for the indoor one, there is a clear probability's increasing. However, drivers for opening and closing actions can be different. For example, in residential buildings CO_2 concentration has been identified as the main stimulus for openings, while the outdoor temperature mainly triggers closing actions.

In parallel, also the *time of the day* and contingent activities influence window operation [7]. Figure 5.3 reports the probability of openings and closings according to different time intervals. The data concern the school environment and refer to case study B (Appendix A). It can be noted that the interactions are more frequent (highlighted with arrows) during the mid-morning break and between lessons with different teachers (lessons usually last for one hour) since students can modify the window status without interrupt lectures and their own attention. Openings and closings occur at the same intervals or are shifted of few time-steps: such finding suggests that users open the windows to refresh indoor air but they suddenly re-close them to avoid an excessive temperature decreasing.

Peculiar *building features*, surrounding and social factors play an undeniable role in windows control too. Façade orientation, overhangs and window type significantly affect the adjustments (e.g., top-hung windows allow openings even during rainfall) [8]. Outdoor noise level and scarce outdoor air quality can bring people to minimize openings to not threaten their own IEQ. In shared spaces, as

Fig. 5.2 Envelopes for window opening probability in relation to indoor and outdoor temperature [7]

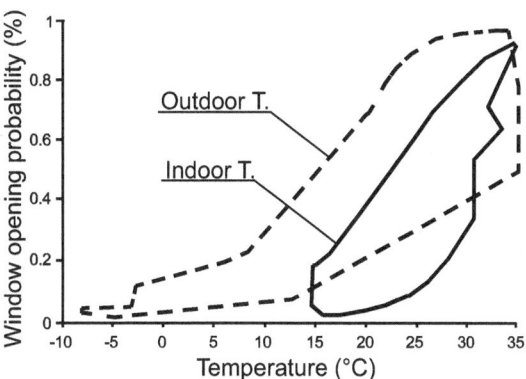

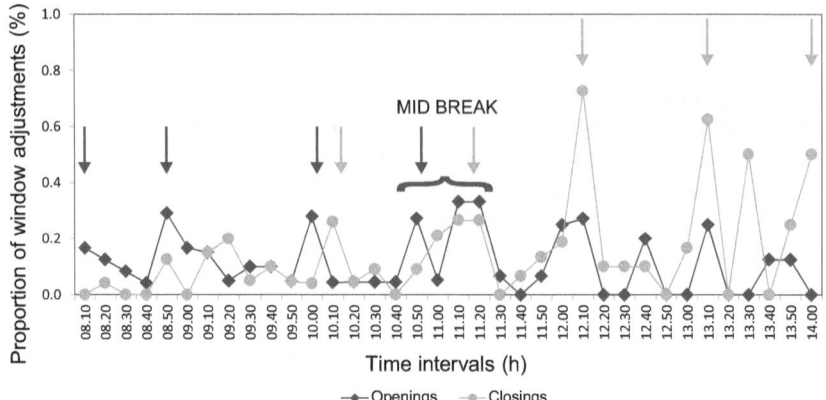

Fig. 5.3 Probability of window opening and closing in relation to time of the day. (Data refer to case study B)

multi-occupied offices, window control can also be a consequence of social rules [9]. In fact, groups of people under similar environmental conditions can behave in dissimilar ways due to different group management and authority (e.g., some individual can impose their preferences because they are the senior members). Even if it is evident the importance of social interactions, such rules can be extremely difficult to represent and, in particular, to extend to different cases study.

Behavioural Models
The high variability in window control encouraged researchers in developing several behavioural models to predict windows status. Most of the models presented in the literature concern office buildings, while dwellings and schools are less described. Environmental and time-related triggers are frequently included. Regarding the office context, other suggested parameters are the previous window state and the user type (i.e., active, medium, passive) [10–13]. To predict windows adjustments in residential buildings, also types of ownership (i.e., owner or rented) and ventilation type (i.e., natural and mechanical) have been suggested as influencing parameters [14, 15].

5.3   Light-Switching

Studied since the seventies, users' interaction with electric lights interested researchers of many different countries. The vast majority of the studies involves office buildings (from single-occupied to open plan offices), while surveys in other building uses are almost missing. Monitoring campaigns concerned different periods and durations but they generally lasted for several months. Researchers assessed that occupants usually prefer working with natural light [16–18], however, when this is lacking, lighting appliances are the main tools to achieve adequate indoor illuminance. In fact, although the lighting use is connected to users' desires and to physiological features, illuminance levels can also be a consequence of specific activities (e.g., drawing), independent from users' preferences. Occupants' and building's management of the lighting system have a direct effect on both electric energy consumptions and on people's visual comfort.

Energy Consumptions
Lighting energy consumptions are mainly related to commercial buildings since they account for 25% of the total energy consumed in the U.S. and 14% in EU [19]. In this perspective, many efforts have been made to design lighting control systems to save energy and reduce costs. Simulation studies, analysing energy reduction through the adoption of occupancy sensors, reported possible energy savings from 20 to 40% [20]. When occupancy sensors are coupled with illuminance ones, a further decreasing (up to 69%) has been reached in comparison to a standard lighting schedule (i.e., lights on for all day) [21].

Comfort Perspective
Visual comfort is of primary importance for occupants' well-being, productivity and satisfaction. It has been widely assessed that glare and low illuminance levels decrease workers satisfaction, while natural light and views make users in a better mood [22]. Moreover, surveys on automated lighting systems highlighted that occupants used to override these systems since they express discomfort and the preference for a tailored indoor climate [23].

Triggering Parameters
Time-related events and environmental variables are both influencing factors for occupants' lighting adjustments. *Time of the day* is one of the main drivers, since switch-on events are focused in the firsts entering the room, while switch-off ones are connected to departures and duration of absences. During intermediate periods lights adjustments are limited. Also *environmental parameters* trigger light-switching behaviours. Among the variables, work-plane illuminance is one of the most used since it properly reflects users' position inside the rooms. Figure 5.4 shows the behavioural models developed for the arrival (black solid line) and intermediate period (grey dashed line). The data, concerning Room 1 of case study A (Appendix A), refer to the heating season (i.e., from November 2016 to April 2017) since most of switching behaviours occur during periods with low outdoor illuminance. It can be noted that at arrival the probability increases significantly for illuminance levels lower than 180 lux and becomes almost certain around 60 lux. During intermediate periods, switching behaviours are much less frequent because occupants tend to adapt to indoor conditions.

Also some *building features* and room properties can significantly affect users' comfort and modify the interaction with the lighting appliances. Densely occupied and open plan offices seem to have a negative effect on visual comfort [24]. Moreover, the orientation of the room, the position of occupants' workstation and lighting controls and the presence of shading devices have been identified as

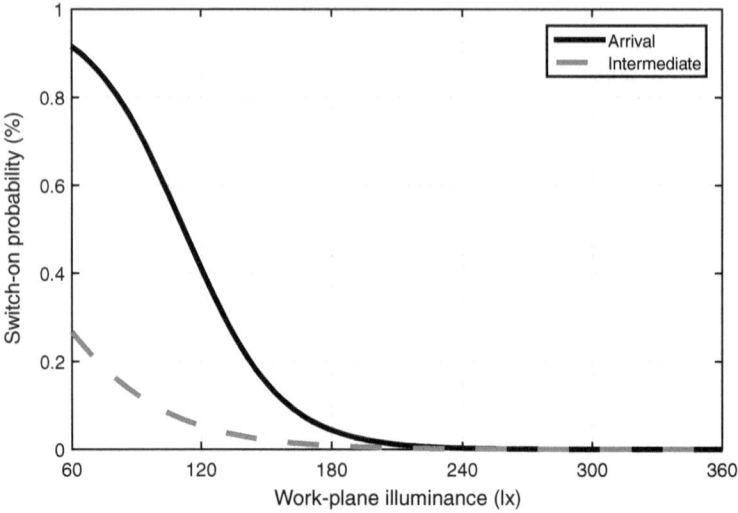

Fig. 5.4 Difference in switch-on probability according to arrival and intermediate periods. (Data refers to case study A)

influencing elements too [25]. In particular, the position of the switches, usually located at the entrance of the room, can be one of the main reasons for the high frequency of turn-on events at arrival.

Behavioural Models
Several behavioural models, that consider the patterns highlighted before, have been proposed and some of them have been also implemented in simulation software. Most of them include at least environmental and time-related variables [26, 27] or occupancy patterns [28–30]. Few papers also found a statistical correlation with thermal variables (e.g., outdoor temperature) [31, 32].

5.4   Shadings and Blinds

The earliest researches on shading and blind usage date back to 1978. This topic has been widely studied, in particular, along USA and Europe. The monitoring campaigns, even if performed only in office buildings, differ a lot for their duration (from few days to several years) and for the involved period (only summer or winter or a random sampling). Several studies also recorded data from an entire building in order to evaluate behaviours according to different exposures. The analyses of this topic and the comparisons between different studies can be laborious and sometimes challenging. In fact, the differences in the building latitude, shading and glazing typology, shading shape and position in relation to the envelope can make a case study so much peculiar to be incomparable to others [33].

Energy Consumptions and Comfort Perspective

The use of shading systems is influenced by users' thermal and visual comfort since the penetration of daylight increase the indoor temperature and the illuminance levels. Consequently, shadings and blinds management turns out to be one of the most influencing pattern on building energy consumptions [34]. A careful use can decrease the required thermal energy in winter and also save electricity, increasing the indoor illuminance. On the contrary, wasteful behaviours can significantly impact on the total energy consumptions.

Triggering Parameters

Several researchers investigated the triggers for shading and blinds adjustments but a general agreement has not been reached jet. Most of the surveys indicated that the *variables related to visual comfort* are the most influencing ones. In particular, illuminance levels (e.g., work plane, external, horizontal) [35–37], solar radiation [38–40] and glare phenomenon [23, 41, 42]. However, also indoor and outdoor temperature have been identified as triggering factors since it was noted an increase in blind use when both the temperatures rise [43].

Also the analyses on dependence from *time-related events* and *season* showed dissimilar conclusions. In fact, while some authors [37] did not note any correlation with time of the day, others reported an increasing frequency of openings on arrivals and departures [44]. Similarly, a major use has been assessed both during wintertime [45] and summer season [40].

Despite the differences among the above-mentioned results, some similar conclusions have been reached. Researchers noted that once the occupants set blinds position, it remains unchanged until a discomfort situation occurs [40]. Whether no triggering event happens, the position can be kept constant also for entire months [16, 38]. Many researchers also observed that users tend to lower the blinds at illuminance levels higher than those at which they open them [36]. This behaviour is called "hysteresis phenomenon".

Also some peculiar *shading's features and applications* affect the users' interaction. In fact, recently, new technologies have been directed to the development and application of dynamic integrated systems which aim at providing shading while optimizing visual comfort and avoiding glare [46]. However, many researchers highlighted that automated systems are usually overwritten by manual adjustments and, as a consequence, motorized blinds can be modified up to three times more frequently than manual ones [1, 23, 47]. In fact, shading use is a consequence of both visual and thermal comfort as well as of outside view or need for privacy. Moreover, occupants are generally more satisfied with the environment when they can customise the systems to their preferences and needs.

Behavioural Models

The numerous surveys and monitoring campaigns promoted the development of several behavioural models. Most of the models aimed at predicting shading/blind use are coupled with light switching models. In fact, several surveys highlighted that such actions are frequently correlated [38, 48, 49] and, especially, during arrivals and departures [44]. They are usually driven by visual parameters (e.g., indoor and outdoor illuminance) and only in few cases decoupled according to time of the day [50]. However, analyses aimed at assessing the best model did not produced robust results. It has been suggested that to improve behavioural models predictions also non-physical parameters (e.g., need for privacy and accessibility to controls) should be included since they play a substantial role in shading/blind adjustments [48].

5.5   Thermostat and Air-Conditioning Use

Studies concerning thermostat and air-conditioning (AC) use and regulation are mainly connected to residential buildings. In fact, at home, people can freely turn on and off the systems and also regulate the set-point temperature. On the contrary, at work users usually have to tolerate the building management and when they feel discomfort, they have to adopt other strategies (e.g., modify the clothing level). Even if it is undeniable that thermostat and air-conditioning usage greatly influences people's thermal comfort and building energy consumptions [51], not many studies focused on this target and most of them have been developed in a limited geographic area (i.e., the Far East). However, the surveys usually covered several months and interested many different buildings or apartments.

Energy Consumptions
The regulation of these systems is probably the most influencing action on building energy consumptions [52]. Researchers found that about half of the energy use in apartment buildings accounts for space heating [53, 54] but it can be greatly connected to dwelling features (primarily the building envelope) and building characteristics (up to 42% of the variation is related to them) [55]. Also household composition, lifestyle and thermal preferences affect consumptions. However, occupants' behaviour impact can increase in relation to the different type of dwelling and HVAC system.

Raising the cooling set-point and lowering the heating one, even of few degrees, can lead to a wide reduction of energy consumptions without threating users' thermal comfort. A simulation study, performed on retrofitted office buildings in seven different climates, found a cooling energy saving of 29%, increasing the cooling set-point from 22.2 to 25 °C, and a reduction of 34% of terminal heating energy, reducing the heating set-point of about 1 degree (i.e., from 21.1 to 20 °C) [56].

Comfort Perspective
Occupants' thermal comfort is strictly connected to thermostat and AC adjustments. As highlighted in previous sections, people with great personal control on the building systems tend to adapt easier (accepting wider ranges of indoor thermal environments) in comparison to users that have to passively accept an imposed management. Researchers found that occupants with free control have a neutral indoor operative temperature lower than 2.6 °C during winter and they also

expressed minor intention to change their current thermal conditions than users without the capability of personal control [57]. However, people's neutral temperature varies a lot among different samples [58]. In fact, summer neutral temperature is ranged about between 23 and 29 °C, while the winter one varies in the interval 20–22.5 °C [59]. It has been suggested that the availability of more comfortable environments in comparison to past years has changed users' perception and shifted their comfort zone towards warmer temperatures in winter and cooler in summer.

Triggering Parameters

The above-mentioned outcomes suggest that users are mainly triggered by *thermal parameters* in adjusting thermostats and AC units. Indoor and outdoor temperatures have been identified as the major stimuli for both these actions. Due to the variability in users' perceptions, household typology (e.g., singles, families) habits and routine [60], the range of switch-on and off behaviours is wide. Figure 5.5 reports the envelopes of many behavioural models developed to predict AC switching behaviours [7]. The figure highlights that the turn-on probability starts for indoor temperature between 25 and 33 °C and outdoor temperature higher than 20 °C [61–63]. Instead, turn-off actions begin in the indoor temperature range 27–30 °C [61]. These actions are also related to a specific *time of the day* since researchers found a substantial difference between working days and weekends [64], and also in relation to times of the day that are connected to specific activities (e.g., sleeping and eating) [61, 65, 66].

Behavioural Models

Behavioural models related to this topic mainly concerned the prediction of the air-conditioning use. Including environmental parameters and time-related events, results showed good agreements with recorded behaviours [61, 63]. However, the target of many studies was the investigation of the impact on energy consumptions of different set-points preferences. Peculiar set-points have been related to different behaviour styles (e.g., waster, energy conscious) and the analyses on the energy use underlined the big influence of users' preferences [67–69].

Fig. 5.5 Envelopes for the probability of AC units' switch-on in relation to indoor and outdoor temperature [7]

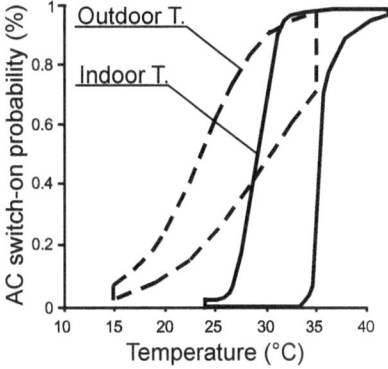

5.6   Fans and Doors

Not many studies focused on the analysis of fans and doors usage. All the surveys have been performed in office buildings and most of them studied these controls in parallel since they can modify users' thermal perception [32] with a very little affection on building energy consumptions [34, 43]. Doors and fans, in particular, are systems that condition the user in a limited region of the body, independently of the HVAC system.

Comfort Perspective
When these devices are adopted in cooperation with others systems (windows in particular), they can improve the indoor air quality and reduce the perceived indoor temperature, due to the increase of ventilation inside the rooms [70]. In fact, surveys found that above the neutral thermal sensation (e.g., feeling warm or hot) more than 50% of people would prefer an increase in the air movement [71]. Personalised cooling systems (e.g., fans) allow expanding users' thermal comfort zone and reducing HVAC energy consumptions at the same time. Surveys [72] found that at the indoor temperature higher than 30 °C thermal comfort can be maintained using air speeds of as much as 1.5 m/s in the upper region of the body while saving up to 60% of HVAC energy. However, personalised cooling cannot compensate discomfort sensations triggered by extreme thermal environments (i.e., ambient temperature of up to 30 °C and relative humidity of up to 70%).

Triggering Parameters
The adoption of fans and doors is driven by a thermal discomfort. As a consequence, indoor and outdoor temperatures have been identified as the triggering factors for their use. The envelopes reported in Fig. 5.6 derived from several behavioural models developed to predict fans use [7]. Starting at about 20 °C for the indoor temperature and at 15 °C for the outdoor one, their employment is almost certain around 30 °C [41, 43, 73–75]. In contrast to the devices analysed in previous sections, no study investigated a correlation with specific events or time of the day.

Fig. 5.6 Envelopes for the fans switch-on in probability in relation to indoor and outdoor temperature [7]

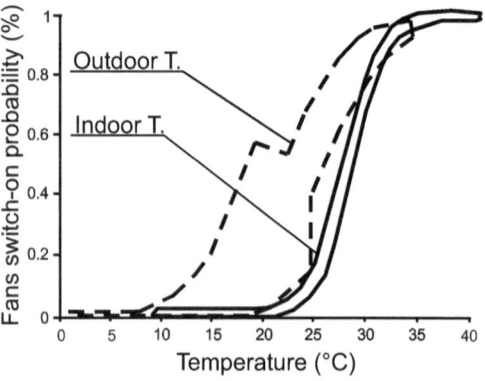

Behavioural Models

Some behavioural models have been proposed [43, 74] but only one [73] has been implemented in a building simulation program to assess the validity of the predictions. Results showed acceptable correspondences but further studies are needed for a better comprehension of these devices. In fact, since the use of fans and doors shifts the normal indoor temperature, it can allow building engineers and managements to adjust the temperature set-points when including them in their designs. Careful evaluations on the enhanced air movement enable substantial energy savings, expanding zone temperature set-points and decreasing HVAC intensity [76].

5.7   Actions' Priority

Very few studies tried to understand the order of use of adaptive adjustments; in fact, this issue remains still unaddressed. Despite this lack, it is not of a secondary importance to evaluate sequences of behaviours since they can have an enormous impact on the energy use (i.e., up to a factor of 3.3) [77].

When people have different possibilities of adaptation it is important to understand which one they are more likely to take. Analyses related to office buildings proposed very different results. One study [34] suggested that thermal discomfort drives the users to adjust first the set-point temperature, then the clothing level and finally blinds position. A different conclusion was proposed by Langevin et al. [78], which indicated the clothing insulation as the main action to restore thermal comfort both in naturally ventilated (NV) and air-conditioning (AC) buildings. Then, the sequence of behaviours diversifies in relation to the building system: in NV buildings, users primarily interact with windows and then with fans, while in AC ones, the sequence is just the opposite. Also doors seem to be opened more frequently when the windows were already open [79].

In offices, occupants are more likely to make, at first, personal adjustments and then interact with most personal (and also nearest) devices. Such behaviour was also noted in energy efficient buildings, in which users tend to prefer personal adjustments to environmental ones to increase their comfort condition [80]. This behaviour can be very effective in an energy saving perspective. In fact, the energy use of personal comfort systems (PCS) is negligible (especially if they are also efficient) compared to the one of building central systems, so it would be useful broadening the range of indoor temperatures through the inclusion of PCS to maintain comfort levels [56].

In residential buildings, it was recorded a different behaviour depending on the thermal sensation [57]. When occupants feel hot they prefer to decrease the heating set-point, while when they feel cold their first action is to add clothing. The authors suggested that these findings could be a consequence of economic issues since a lower space heating temperature will result in lower heating fees.

The envelopes previously reported in Fig. 5.2, 5.5 and 5.6 and derived from a comprehensive literature review on building devices [7] have been combined in

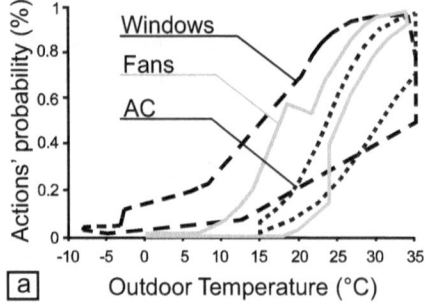 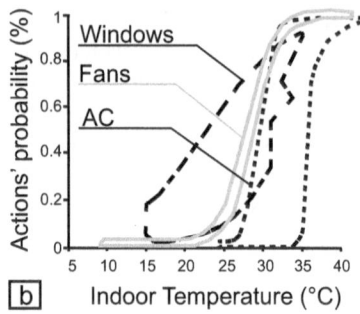

Fig. 5.7 The envelopes report the actions' probability for several actions in relation to both outdoor and indoor temperature [7]

Fig. 5.7. The Figure suggests that the sequence windows-fans-AC could be the most commonly adopted. However, among the studies are present big differences related to buildings' features, systems and also to users' control on the devices. So, the sequence of behaviours remains still an open question [75].

References

1. Belafi Z, Hong T, Reith A (2017) Smart building management versus intuitive human control: lessons learnt from an office building in Hungary. Build Simul. https://doi.org/10.1007/s12273-017-0361-4
2. Sorgato MJ, Melo AP, Lamberts R (2016) The effect of window opening ventilation control on residential building energy consumption. Energy Build 133:1–13. https://doi.org/10.1016/j.enbuild.2016.09.059
3. Wang L, Greenberg S (2015) Window operation and impacts on building energy consumption. Energy Build 92:313–321. https://doi.org/10.1016/j.enbuild.2015.01.060
4. Yousefi F, Gholipour Y, Yan W (2017) A study of the impact of occupant behaviors on energy performance of building envelopes using occupants' data. Energy Build 148:182–198. https://doi.org/10.1016/j.enbuild.2017.04.085
5. Pan S, Xu C, Wei S et al (2016) Improper window use in office buildings: findings from a longitudinal study in Beijing, China. Energy Procedia 88:761–767. https://doi.org/10.1016/j.egypro.2016.06.104
6. Borgeson S, Brager G (2008) Occupant control of windows: accounting for human behavior in building simulation
7. Stazi F, Naspi F, D'Orazio M (2017) A literature review on driving factors and contextual events influencing occupants' behaviours in buildings. Build Environ 118:40–66. https://doi.org/10.1016/j.buildenv.2017.03.021
8. Roetzel A, Tsangrassoulis A, Dietrich U, Busching S (2010) A review of occupant control on natural ventilation. Renew Sustain Energy Rev 14:1001–1013. https://doi.org/10.1016/j.rser.2009.11.005
9. Fabi V, Andersen RV, Corgnati S, Olesen BW (2012) Occupants' window opening behaviour: a literature review of factors influencing occupant behaviour and models. Build Environ 58:188–198. https://doi.org/10.1016/j.buildenv.2012.07.009

10. Herkel S, Knapp U, Pfafferott J (2008) Towards a model of user behaviour regarding the manual control of windows in office buildings. Build Environ 43:588–600. https://doi.org/10.1016/j.buildenv.2006.06.031

11. Yun GY, Steemers K (2008) Time-dependent occupant behaviour models of window control in summer. Build Environ 43:1471–1482. https://doi.org/10.1016/j.buildenv.2007.08.001

12. Yun GY, Tuohy P, Steemers K (2009) Thermal performance of a naturally ventilated building using a combined algorithm of probabilistic occupant behaviour and deterministic heat and mass balance models. Energy Build 41:489–499. https://doi.org/10.1016/j.enbuild.2008.11.013

13. Rijal HB, Tuohy PG, Humphreys M et al (2007) Using results from field surveys to predict the effect of open windows on thermal comfort and energy use in buildings. Energy Build 39:823–836. https://doi.org/10.1016/j.enbuild.2007.02.003

14. Fabi V, Andersen RV, Corgnati SP (2012) Window opening behaviour : simulations of occupant behaviour in residential buildings using models based on a field survey. In: 7th Windsor conference: the changing context of comfort an unpredictable world, pp 12–15

15. D'Oca S, Fabi V, Corgnati SP, Andersen RK (2014) Effect of thermostat and window opening occupant behavior models on energy use in homes. Build Simul 7:683–694. https://doi.org/10.1007/s12273-014-0191-6

16. Galasiu AD, Veitch JA (2006) Occupant preferences and satisfaction with the luminous environment and control systems in daylit offices: a literature review. Energy Build 38:728–742. https://doi.org/10.1016/j.enbuild.2006.03.001

17. Veitch JA, Gifford R (1996) Assessing beliefs about lighting effects on health, performance, mood, and social behavior. Environ Behav 25:446–470. https://doi.org/10.1177/0013916596284002

18. Inoue T, Kawase T, Ibamoto T et al (1988) The development of an optimal control system for window shading devices based on investigations in office buildings. In: ASHRAE Transactions American Society of Heating, Refrigerating and Air-conditioning Engineers, pp 1034–1049

19. IEA (2010) Guidebook on energy efficient electric lighting for buildings in Annex 45-Energy efficient electric lighting for buildings. Espoo, Finland

20. Nagy Z, Yong FY, Frei M, Schlueter A (2015) Occupant centered lighting control for comfort and energy efficient building operation. Energy Build 94:100–108. https://doi.org/10.1016/j.enbuild.2015.02.053

21. Chiogna M, Mahdavi A, Albatici R, Frattari A (2012) Energy efficiency of alternative lighting control systems. Light Res Technol 44:397–415

22. Serghides DK, Chatzinikola CK, Katafygiotou MC (2015) Comparative studies of the occupants' behaviour in a university building during winter and summer time. Int J Sustain Energy 34:528–551. https://doi.org/10.1080/14786451.2014.905578

23. Reinhart CF, Voss K (2003) Monitoring manual control of electric lighting and blinds. Light Res Technol 35:243–258. https://doi.org/10.1191/1365782803li064oa

24. Al horr Y, Arif M, Katafygiotou M et al (2016) Impact of indoor environmental quality on occupant well-being and comfort: a review of the literature. Int J Sustain Built Environ 5:1–11. https://doi.org/10.1016/j.ijsbe.2016.03.006

25. Fabi V, Andersen RK, Corgnati S (2016) Accounting for the uncertainty related to building occupants with regards to visual comfort : a literature survey on drivers and models. Buildings. https://doi.org/10.3390/buildings6010005

26. Hunt DRG (1979) The use of artificial lighting in relation to daylight levels and occupancy. Build Environ 14:21–33. https://doi.org/10.1016/0360-1323(79)90025-8

27. Reinhart CF (2004) Lightswitch-2002: a model for manual and automated control of electric lighting and blinds. Sol Energy 77:15–28. https://doi.org/10.1016/j.solener.2004.04.003

28. Pigg S, Eilers M, Consultant I et al (1996) Behavioral aspects of lighting and occupancy sensors in privates offices: a case study of a University Office Building. ACEEE 1996 Summer Study Energy Effic Build 161–170

29. Moore T, Carter DJ, Slater A (2003) Long-term patterns of use of occupant controlled office lighting. Light Res Technol 35:43–59. https://doi.org/10.1191/1477153503li061oa
30. Yun GY, Kong HJ, Kim H, Kim JT (2012) A field survey of visual comfort and lighting energy consumption in open plan offices. Energy Build 46:146–151. https://doi.org/10.1016/j.enbuild.2011.10.035
31. Andersen RV, Toftum J, Andersen KK, Olesen BW (2009) Survey of occupant behaviour and control of indoor environment in Danish dwellings. Energy Build 41:11–16. https://doi.org/10.1016/j.enbuild.2008.07.004
32. Nicol JF (2001) Characterising occupant behavior in buildings: towards a stochastic model of occupant use of windows, lights, blinds heaters and fans. In: 7th International IBPSA Conference, pp 1073–1078
33. Stazi F, Marinelli S, Di Perna C, Munafò P (2014) Comparison on solar shadings: monitoring of the thermo-physical behaviour, assessment of the energy saving, thermal comfort, natural lighting and environmental impact. Sol Energy 105:512–528. https://doi.org/10.1016/j.solener.2014.04.005
34. Bonte M, Thellier F, Lartigue B (2014) Impact of occupant's actions on energy building performance and thermal sensation. Energy Build 76:219–227. https://doi.org/10.1016/j.enbuild.2014.02.068
35. Nicol F, Wilson M, Chiancarella C (2006) Using field measurements of desktop illuminance in European offices to investigate its dependence on outdoor conditions and its effect on occupant satisfaction, and the use of lights and blinds. Energy Build 38:802–813. https://doi.org/10.1016/j.enbuild.2006.03.014
36. Sutter Y, Dumortier D, Fontoynont M (2006) The use of shading systems in VDU task offices: a pilot study. Energy Build 38:780–789. https://doi.org/10.1016/j.enbuild.2006.03.010
37. Haldi F, Robinson D (2009) A comprehensive stochastic model of blind usage : theory and validation. In: Building simulation conference Glasgow, Scotland, pp 545–552
38. Gunay HB, O'Brien W, Beausoleil-Morrison I, Gilani S (2016) Development and implementation of an adaptive lighting and blinds control algorithm. Build Environ 113: http://dx.doi.org/10.1016/j.buildenv.2016.08.027
39. Yao J (2014) Determining the energy performance of manually controlled solar shades: a stochastic model based co-simulation analysis. Appl Energy 127:64–80. https://doi.org/10.1016/j.apenergy.2014.04.046
40. Zhang Y, Barrett P (2012) Factors influencing occupants' blind-control behaviour in a naturally ventilated office building. Build Environ 54:137–147. https://doi.org/10.1016/j.buildenv.2012.02.016
41. Raja IA, Nicol JF, McCartney KJ, Humphreys MA (2001) Thermal comfort: use of controls in naturally ventilated buildings. Energy Build 33:235–244. https://doi.org/10.1016/S0378-7788(00)00087-6
42. Maniccia D, Rutledge B, Rea M, Morow W (1999) Occupant use of manual lighting controls in private offices. Iesna 489–512. https://doi.org/10.1080/00994480.1999.10748274
43. Haldi F, Robinson D (2008) On the behaviour and adaptation of office occupants. Build Environ 43:2163–2177. https://doi.org/10.1016/j.buildenv.2008.01.003
44. Correia da Silva P, Leal V, Andersen M (2013) Occupants interaction with electric lighting and shading systems in real single-occupied offices: results from a monitoring campaign. Build Environ 64:152–168. https://doi.org/10.1016/j.buildenv.2013.03.015
45. Rubin AI, Collins BL, Tibbott RL (1978) Window blinds as a potential energy saver-a case study. Natiolal Bureau of Standards, Washington DC
46. Konstantoglou M, Tsangrassoulis A (2016) Dynamic operation of daylighting and shading systems: a literature review. Renew Sustain Energy Rev 60:268–283. https://doi.org/10.1016/j.rser.2015.12.246
47. Vine E, Lee E, Clear R et al (1998) Office worker response to an automated Venetian blind and electric lighting system: a pilot study. Energy Build 28:205–218. https://doi.org/10.1016/S0378-7788(98)00023-1

48. Sadeghi SA, Karava P, Konstantzos I, Tzempelikos A (2016) Occupant interactions with shading and lighting systems using different control interfaces: a pilot field study. Build Environ 97:177–195. https://doi.org/10.1016/j.buildenv.2015.12.008

49. Sadeghi SA, Awalgaonkar NM, Karava P, Bilionis I (2017) A Bayesian modeling approach of human interactions with shading and electric lighting systems in private offices. Energy Build 134:185–201. https://doi.org/10.1016/j.enbuild.2016.10.046

50. Haldi F, Robinson D (2010) Adaptive actions on shading devices in response to local visual stimuli. J Build Perform Simul 3:135–153

51. EC (2004) European Union energy and transport in Figures, 2006 ed. Part 2, Dictatorate General for Energy and Transport, European Commission, Brussels

52. Andersen R, Olesen B, Toftum J (2011) Modelling occupants' heating set-point prefferences. In: 12th Conference of International Building Performance Simulation Association. Sydney, 14–16 November, pp 151–156

53. Engvall K, Lampa E, Levin P et al (2014) Interaction between building design, management, household and individual factors in relation to energy use for space heating in apartment buildings. Energy Build 81:457–465. https://doi.org/10.1016/j.enbuild.2014.06.051

54. O'Neill Z, Niu F (2017) Uncertainty and sensitivity analysis of spatio-temporal occupant behaviors on residential building energy usage utilizing Karhunen-Loève expansion. Build Environ 115:157–172. https://doi.org/10.1016/j.buildenv.2017.01.025

55. Guerra Santin O, Itard L, Visscher H (2009) The effect of occupancy and building characteristics on energy use for space and water heating in Dutch residential stock. Energy Build 41:1223–1232. https://doi.org/10.1016/j.enbuild.2009.07.002

56. Hoyt T, Arens E, Zhang H (2014) Extending air temperature setpoints: simulated energy savings and design considerations for new and retrofit buildings. Build Environ 88:89–96. https://doi.org/10.1016/j.buildenv.2014.09.010

57. Luo M, Cao B, Zhou X et al (2014) Can personal control influence human thermal comfort? a field study in residential buildings in China in winter. Energy Build 72:411–418. https://doi.org/10.1016/j.enbuild.2013.12.057

58. Humphreys MA, Hancock M (2007) Do people like to feel "neutral"? Energy Build 39:867–874. https://doi.org/10.1016/j.enbuild.2007.02.014

59. Bae C, Chun C (2009) Research on seasonal indoor thermal environment and residents' control behavior of cooling and heating systems in Korea. Build Environ 44:2300–2307. https://doi.org/10.1016/j.buildenv.2009.04.003

60. Xu B, Fu L, Di H (2009) Field investigation on consumer behavior and hydraulic performance of a district heating system in Tianjin, China. Build Environ 44:249–259. https://doi.org/10.1016/j.buildenv.2008.03.002

61. Ren X, Yan D, Wang C (2014) Air-conditioning usage conditional probability model for residential buildings. Build Environ 81:172–182. http://dx.doi.org/10.1016/j.buildenv.2014.06.022

62. Tanimoto J, Hagishima A (2005) State transition probability for the Markov Model dealing with on/off cooling schedule in dwellings. Energy Build 37:181–187. https://doi.org/10.1016/j.enbuild.2004.02.002

63. Schweiker M, Shukuya M (2009) Comparison of theoretical and statistical models of air-conditioning-unit usage behaviour in a residential setting under Japanese climatic conditions. Build Environ 44:2137–2149. https://doi.org/10.1016/j.buildenv.2009.03.004

64. Kempton W, Feuermann D, McGarity AE (1992) "I always turn it on super": user decisions about when and how to operate room air conditioners. Energy Build 18:177–191. https://doi.org/10.1016/0378-7788(92)90012-6

65. Habara H, Yasue R, Shimoda Y (2013) Survey on the Occupant Behavior Relating to Window and Air Conditioner Operation in the Residential Buildings. In: 13th Conference of International Building Performance Simulation Association, pp 2007–2013

66. Lin B, Wang Z, Liu Y et al (2016) Investigation of winter indoor thermal environment and heating demand of urban residential buildings in China's hot summer—cold winter climate region. Build Environ 101:9–18. https://doi.org/10.1016/j.buildenv.2016.02.022

67. Sun K, Hong T (2017) A framework for quantifying the impact of occupant behavior on energy savings of energy conservation measures. Energy Build 146:383–396. https://doi.org/10.1016/j.enbuild.2017.04.065
68. Hong T, Lin HW (2013) Occupant Behavior: impact on energy use of private offices. Lawrence Berkeley Natl. Lab. Berkeley, CA
69. Peng C, Yan D, Wu R et al (2012) Quantitative description and simulation of human behavior in residential buildings. Build Simul 5:85–94. https://doi.org/10.1007/s12273-011-0049-0
70. Wallace LA, Emmerich SJ, Howard-Reed C (2002) Continuous measurements of air change rates in an occupied house for 1 year: the effect of temperature, wind, fans, and windows. J Expo Anal Environ Epidemiol 12:296–306. https://doi.org/10.1038/sj.jea.7500229
71. Zhang H, Arens E, Fard SA et al (2007) Air movement preferences observed in office buildings. Int J Biometeorol 51:349–360. https://doi.org/10.1007/s00484-006-0079-y
72. Veselý M, Zeiler W (2014) Personalized conditioning and its impact on thermal comfort and energy performance—a review. Renew Sustain Energy Rev 34:401–408. https://doi.org/10.1016/j.rser.2014.03.024
73. Rijal HB, Tuohy PG, Nicol JF et al (2008) Development of adaptive algorithms for the operation of windows, fans and doors to predict thermal comfort and energy use in Pakistani buildings. ASHRAE Trans 114:555–573
74. Nicol JF (2001) Characterising occupant behavior in buildings: towards a stochastic model of occupant use of windows, lights, blinds heaters and fans. In: 7th International IBPSA Conference, pp 1073–1078
75. Rijal HB, Tuohy P, Humphreys MA et al (2011) An algorithm to represent occupant use of windows and fans including situation-specific motivations and constraints. Build Simul 4:117–134. https://doi.org/10.1007/s12273-011-0037-4
76. Zhang H, Arens E, Zhai Y (2015) A review of the corrective power of personal comfort systems in non-neutral ambient environments. Build Environ 91:15–41. https://doi.org/10.1016/j.buildenv.2015.03.013
77. Andersen R (2009) Occupant behaviour with regard to control of the indoor environment
78. Langevin J, Gurian PL, Wen J (2015) Tracking the human-building interaction: a longitudinal field study of occupant behavior in air-conditioned offices. J Environ Psychol 42:94–115. https://doi.org/10.1016/j.jenvp.2015.01.007
79. Haldi F, Robinson D (2011) Modelling occupants' personal characteristics for thermal comfort prediction. Int J Biometeorol 55:681–694. https://doi.org/10.1007/s00484-010-0383-4
80. Azizi NSM, Wilkinson S, Fassman E (2015) An analysis of occupants response to thermal discomfort in green and conventional buildings in New Zealand. Energy Build 104:191–198. https://doi.org/10.1016/j.enbuild.2015.07.012

Chapter 6
Experimental Data Acquisition

Abstract The acquisition of people's behaviours in buildings is extremely important for the evaluation and the improvement of the building performances, as well as for the development of behavioural predictive models. Both occupancy patterns and users' interactions with devices influence the building energy use. Therefore, researchers adopted many techniques to record these data, customising the monitoring system according to the building features and the research aim. This Chapter offers an overview of the techniques adopted in experimental campaigns to detect occupancy patterns and acquire behavioural and environmental data.

6.1 Introduction

The occupants' perspective has become one of the main focuses in the building sector. People's lifestyle can be studied both in existing buildings and in laboratories but the former context is the preferred one since behaviours are more "real" than in the latter. Many techniques have been proposed and adopted to record users' presence and behaviours inside buildings but a standardised method is still missing. This lack allows researchers to employ many different strategies which, in turn, increase the difficulties in comparisons between different results. The current Chapter focuses on the tools and appliances used to monitor both occupancy (Sect. 6.2) and behavioural (Sect. 6.3) data. In particular, the latest and newest technologies are illustrated to give useful suggestions for the development of a monitoring campaign.

6.2 Occupancy Data

Knowing whether a space is occupied or empty is of primary importance both for the development of behavioural models and for the building management optimisation. In fact, several studies [1–3] highlighted that most of the energy use occurred during nonworking hours, due, especially, to inaccurate use of plug-loads.

© The Author(s) 2018   47
F. Stazi and F. Naspi, *Impact of Occupants' Behaviour on Zero-Energy Buildings*,
SpringerBriefs in Energy, https://doi.org/10.1007/978-3-319-71867-5_6

Many techniques have been adopted to identify people's presence and one of the most used concerns passive infrared sensors (PIR) [4, 5]. Such technology is useful for counting people's passages since it recognises the human motion. PIR are widely adopted because they are low-cost and few intrusive. However, the number of persons in the space cannot be recorded as well as immobile occupants (e.g., people working at the desk in offices). PIR can be used to manage the building systems, for example to switch off lights in empty rooms. In such contexts, the delay value assumes a central importance since the system can detect false absences (e.g., static people). Long delays are usually preferred to reduce users' annoyance but they greatly diminish the possible energy savings [6].

PIR can be coupled with CO_2 sensors [7] which clearly assess whether a room is occupied [8]. This union is useful to improve detection accuracy since the creation of a sensor network integrates different types of technology. In fact, the use of CO_2 sensors alone was discouraged because the CO_2 distribution can be easily biased by ventilation rates [6].

Such limitations can be overcome by using video cameras which clearly detect occupancy and number of occupants. However, this method is quite costly both for the camera installation and the data analysis and, in addition, causes privacy issues [9].

Recently, many applications adopted advanced technology to optimise occupancy detection. Bluetooth technology [10], which is provided by the majority of digital devices, is an accurate and low-cost solution for short-range wireless communication. Researchers adopting this technique to estimate positioning provided good results, especially when gradient filters were applied.

Similarly, Radio-frequency identification (RFID) tags track each person continuously throughout the building [18]. A RFID system usually consists of two components (i.e., the reader and the tag) which operate at a certain frequency. When attached to an object, the tag stores a specific ID allowing a personal identification. Combinations of different tags, readers, and frequencies of communication offer a wide flexibility in customizing RFID systems [11]. RFID tags have been also used coupled with PIR sensors to control lighting in multi-occupant offices [12]. This method provided an accuracy of about 91% and so, the authors suggested adopting it to monitor also building energy, use and security.

An improvement has been reached through Wi-Fi based technology [13]. This is a low-cost and simply-implemented technology, which can detect the position and the number of occupants in real-time. Moreover, almost all mobile devices are provided with Wi-Fi, making its application easy also at a large scale. Researcher testing this approach provided a very detailed visualisation and analysis of occupancy patterns, also including spatial distributions and temporal variations [13].

Table 6.1 describes the techniques adopted in previous studies to detect people's presence in buildings. The technologies are listed from the most simple (i.e., manual detection) to the latest and most innovative ones. The second column highlights the main characteristics and issues of each approach, while the last column reports the reference papers.

Table 6.1 Identification of the approaches adopted to detect occupancy in buildings

Instrument	Features and issues	References
Diaries	– Extremely low-cost; – Detection is usually not precise; – Time-consuming for filling and analysis; – Very difficult to manage since recordings are demanded only to users willingness	[14]
Motions detectors (passive infrared and ultrasonic)	– They can identify occupancy transitions; – Nonintrusive; – Low-cost; – Unable to detect nearly motionless occupants (offices and sedentary activities in home)	[1, 9] [4, 5] [15, 16] [17, 18]
Carbon dioxide sensors	– Identify room occupancy; – CO_2 trend can be modified by ventilation, room geometry, windows and doors position; – CO_2 decay is not immediate, short absences can be not detected	[6, 8]
Motion detectors coupled with carbon dioxide sensors	– Improve occupancy detection accuracy; – The significant delay between occupancy and CO_2 increase makes their use inappropriate in open spaces	[7, 19] [20]
Video cameras with computer vision	– Occupancy and number of occupants are monitored simultaneously; – Installation and data analysis are costly; – Intrusive; – Privacy implications	[8, 9] [21]
Wearable sensors	– Precise detection; – Possibility to acquire further data (e.g., met); – Some of them could be extremely intrusive; – If wireless sensors are embedded in clothing or accessories interferences are limited (e.g., smart watches); – Acquisitions for long periods are challenging; – Privacy implications	[22, 23] [24]
Mobile devices	– Smart phones and Wi-Fi hotspots are widely available; – Clear detection of every single occupant; – No data if smart phones are powered off or connected to another Wi-Fi hotspot	[25]
Bluetooth based indoor positioning	– Detection of position and number of occupants in real-time; – Low cost; – Lower power consumption; – Short-range communications	[10]
Radio-frequency identification (RFID) tags	– Track occupants continuously; – Mainly adopted in offices	[11, 12] [26]
Wi-Fi based indoor positioning	– Detection of position and number of occupants in real-time; – Low cost; – Privacy friendly	[13]

6.3   Environmental and Behavioural Data

The first step in the definition of a monitoring campaign concerns the assessment of the study objective since different aims require different instruments and accuracy [27]. For example, the development of a behavioural model requires a high level of detail, while a general estimate of the building energy use needs a lower level of accuracy [1].

The building use typology fixes some limitations too. For example, there are differences between domestic and non-domestic spaces [27]. In particular, monitoring campaigns in public spaces (e.g., offices) are usually less complicated than those in private contexts (e.g., houses). In fact, in addition to strict privacy implications, behaviours in residential buildings offer wide characteristics of occupants, possibilities of activities and devices' controls [28].

Another issue concern the users' knowledge of participating in an experimental survey. In fact, to obtain more reliable results, the subjects should not be aware to be surveyed, but this is almost always not possible because of privacy issues. Also, researchers identified the "Hawthorne effect" (also called the "observer effect") which states that users tend to modify their behaviours because they are informed of being detected [29–31]. Moreover, the installation of sensors can limit the use of some devices and induce occupants to behave in a different way [32].

Despite some restraints, data recording is necessary to investigate how people behave. Acquisition of occupants' behaviours can be performed using objective measurement (through dedicated monitoring systems) and/or subjective ones (surveys, questionnaires). A cooperation of both these methods is the most useful approach since the users' decision-making process is a complicated mechanism [32]. In fact, the recording of objective data is essential for a systematic data acquisition at a fixed time-step and for a mathematical development but, in parallel, questionnaires and individual interviews allow a better understanding of the users' behaviours [33].

Objective Measurements of Behaviours
The simplest and cheapest way to record interactions with building devices concerns the use of daily logs and diaries. Each occupant or a reference occupant for space (e.g., a multi occupied office) should compile a pre-determined format (paper-based or online) for all the investigated actions. Such methods should be more reliable than surveys since occupants record their actions on real-time. However, diaries can be useful only when applied for short periods (e.g., few days) [27]. In fact, if the surveyed period increased, there would be a high probability that subjects drop-off from the study.

Sensors-based technology provides efficiency and accuracy in data acquisition and it also allows monitoring many spaces and behaviours at the same time and for long periods. Such recordings can be made through stand-alone sensors or networks. Even if the first typology is usually low-cost, few intrusive and adaptable to many different contexts, it requires an individual control for each sensor, a frequent download of the data (due to the limited memory) and battery load, if the sensors are not connected to the grid, or, conversely, the necessity of cabling [27]. On the contrary, sensor networks (both wired and wireless) are remotely connected and they usually store the data in the cloud, avoiding issues due to memory space and allowing the data download from any computer. Moreover, building automation/ management systems (BAS/BMS), which provide a centralised management and monitoring of multiple spaces, are becoming more and more popular, especially in new and retrofitted buildings [34–36]. This technology also allows the direct storage of users' actions on the connected devices [16, 37]. However, also such technology presents some issues. In particular, the high costs for initial installation, operation, and maintenance and, sometimes, the difficulty in integrating further sensors with existing BMS and in respecting regulations' norm (e.g., place the sensors at the prescribed height) [38].

Independently from the network typology, the researchers/designers have to choose whether to use wired or wireless technologies. Both of them present advantages and disadvantages. While wireless technologies are usually preferred for their cheap price and easiness in installation (cables are not needed), wired technologies give more security in the data transfer [39]. It is evident that the choice depends on the aims of the monitoring.

Windows and *doors* position is generally recorded using contact or proximity sensors [40–44]. They have the utility to be low-cost and widely diffused but the data are binary saved and so, it is not possible to detect intermediate positions. Recordings through camera-based techniques are more accurate and precise; in fact, Bourikas et al. [45] found an agreement of 97 and 90% for the summer and winter periods, respectively. Table 6.2 summarises the techniques usually adopted to record users' interaction with windows and doors.

Also *shades* and *blinds* position is usually acquired employing camera-based [37] or photographic approaches [16, 51, 52]. Even if this method grants recording lot of observations, it also presents some disadvantages. In particular, outdoor phenomena (e.g., rain, insufficient daylight) can drastically limit the quality and reliability of the acquired data, since cameras are placed in external positioning [53]. Moreover, the technique is expensive and labour intensive, even if computer vision programs can facilitate the work. On the contrary, sensors applied directly on the devices (e.g., accelerometers) [209] provide precise data for long periods but their application is usually limited to few objects [53]. The above-mentioned approaches, also including the peculiar features and issues, are reported in Table 6.3.

The use of *electric lights* is frequently detected placing dedicated sensors in the proximity of the light source [55, 60, 61]. However, this type of interaction can be easily recorded thank to BMS [16, 37]. Table 6.4 presents the most frequent methods reported in previous studies.

Similarly to electric lights, also *thermostat* and *air-conditioning* settings in offices can be acquired through BMS. In domestic contexts, different techniques can be adopted. The heating system operation can be detected through the new

Table 6.2 Identification of the techniques adopted to record windows and doors status

Equipment	Features and issues	References
Contact/proximity sensors	– Low-cost; – Binary state; – Cabling; – Generally not practical for retrofit	[8, 42] [41, 44] [40, 43] [46]
Three-axis accelerometers	– Can determine the opening angle; – Real-time acquisition for wireless sensors; – Easiness in retrofitted contexts; – Cost issues for deployment at scale	[47]
Photographic and camera-based approaches	– The position of hinged windows is more difficult to visually assess because angular adjustment translates to only small apparent movements if the camera is aimed perpendicularly to the façade; – Can determine the opening angle comparing the window frame; – Camera angle to the façade should be at 45 °C – Not economic; – Labour intensive; – Environmental phenomena can influence the status identification (e.g., rain and daylight) – The low resolution to avoid privacy implications decreases the data quality – Limitations on the interpretation of data at large scales and long-term monitoring campaigns	[48, 49]
Surveys and questionnaires	– Actions are not continuously recorded; – Acquisitions are demanded to users' memory	[50]

Table 6.3 Identification of the techniques adopted to record shadings and blinds position

Equipment	Features and issues	References
Electro-mechanical sensors	– Efficiency; – Possibility to record slat angles and deployment level; – Large-scale installations are impractical	[53, 54]
Photographic and camera-based approaches	– Positioned outdoor and facing the building; – Limited to hours of the day in which there is sufficient daylight; – Not economic; – Labour intensive; – The low resolution to avoid privacy implications decreases the data quality; – With Venetian blinds, slat angles are often not discernible from photographs; – Limitations on the interpretation of data at large scales and long-term monitoring campaigns but computer vision programs can simplify the work	[16, 37, 51, 52, 55, 56] [57, 58] [59]
Questionnaires	– Actions are not continuously recorded; – Acquisitions are demanded to users' memory	[50]

Table 6.4 Identification of the techniques adopted to record light-switching behaviours

Equipment	Features and issues	References
Light sensors	– Sensors should be placed near to electric light sources and not covered	[55, 60] [61]
Techniques of pattern recognition	– Particularly reliable in spaces lit by direct electric lighting; – Limitations occur in environments with indirect electric lighting; – Data analysis is time-consuming	[51]
Photographic approaches	– Cameras positioned indoor are reliable; – Data analysis is time-consuming	[62]

generation of digital thermostats, which usually provide remote controls and set-point adjustments storage [63, 64]. The use of the AC units has been mainly recorded installing unit power meter [65–67] or temperature sensors [44, 68] placed at the outlet of the split. Both of them allow a continuous monitoring of the unit but the first one only gives indications on the on/off use and not on the set-point temperature. The techniques used to detect thermostats and AC settings are illustrated in Table 6.5.

Surveys are frequently adopted to investigate the use of *fans* [33, 50, 71] and the users' *clothing level* [72, 73]. This method does not permit continuous observations and it is quite intrusive, especially in relation to occupants' clothing. To overcome such issues, some researchers disaggregated the electrical loads to investigate fan use [74] and used observations for the clothing levels [33, 75, 76].

The metabolic rate, together with clothing level, is a very important data to evaluate users' thermal comfort. Most of the studies refer to tabular values of ISO 9920 [77] or Annex C of ISO 7730 [78] to define clothing insulation and to ISO 8996 [79] or Annex B of ISO 7730 [78] to set the metabolic activity. While evaluations on clothing level are limited to surveys and observations, the recent

Table 6.5 Identification of the techniques adopted to record interactions with thermostats and AC units

Device	Equipment	Features and issues	References
Thermostat	Integrated sensors or set-point logs	– Digital thermostats may include logging capabilities and remote access via the Internet	[63, 64]
	Surveys	– The resolution of the data required for statistical models is not provided	[69, 70]
Air-Conditioning	Unit power meter	– Recorded the on/off state and run time; – Continuous monitoring	[65, 66, 67]
	Temperature sensor	– Placed at the outlet of the split; – Recording of temperature fluctuations	[44, 68]

Table 6.6 Identification of the techniques adopted to record actions on fans, clothing levels and metabolic rates

Device	Equipment	Features and issues	References
Fans	Surveys	– Actions are not continuously recorded; – Acquisitions are demanded to users' memory	[33, 50] [71]
	Disaggregating electrical loads	-Unable to disaggregate between similar appliances.	[74]
Clothing level	Observation	– Not allowed ascertain undergarments without asking; – Small adjustments are difficultly detected	[33, 75] [76]
	Surveys	– Intrusive	[72, 73]
Metabolic rate	Medical diagnostic station	– Accuracy; – Data are not continuous; – Intrusiveness	[82]
	Smart wearable devices	– Smart devices (e.g., watches) has considerably reduced intrusiveness; – Need to know the heart rate for every subject at rest	[80, 81]

development of smart wearable devices (e.g., smart watches) allowed the continuous monitoring of subjects' physical data [80, 81]. In addition, such sensors are low-cost and considerably reduce the intrusiveness in users' life. The Table 6.6 recaps the common methods adopted to investigate the use of fans, the users' clothing levels and metabolic rates.

Subjective Measurements of Behaviours

Subjective measurements are usually adopted to enrich objective findings, investigating aspects undetectable through objective techniques. Questionnaires (both paper and web-based) [83, 84] and interviews [85, 86] are the most common approaches. Using the first method, occupants answer to pre-determined questions with a precise frequency (e.g., once a day) [34, 87], while the second approach is more invasive and time-consuming but it allows deeper investigations on users' attitudes and behaviours [69]. Some studies [88] combine both these methods to improve people's behaviours understanding.

Subjective methods overcome most of the issues referred to objective ones but they have limitations too. Recording the observations manually has big restraints on the sample's dimension, on the monitoring duration and on the overall accuracy [46]. Usually, people get tired of filling in questionnaires and reporting behaviours and actions over time, especially when the requested frequency is high. Moreover, researchers sometimes highlighted inconsistency between actual and self-reported behaviours [31].

Although some limitations exist, this kind of approach should be a valid support to objective measurements since it can provide personal information, preferences and habits about the sample. In fact, the understanding of behaviours' motivations, group interactions, social and contextual correlations can be achieved only through subjective methods.

Environmental Data Acquisition

Many reviews [53, 89, 90] highlighted that environmental data are one of the main factors which stimulate occupants' interactions with building devices; as a consequence, their acquisition is absolutely necessary.

Once the variables to be recorded are selected, probes need to be positioned according to regulations (e.g., ISO 7726:2001 [91]) which define the position and the height of the sensors. Ambient parameters (e.g., indoor temperature and humidity) are generally acquired at the centre of the room (if of modest size) since such position should represent the entire environment quite well. Punctual measurements (e.g., wall surface temperatures) are less useful to assess the human-building interaction. Probes should also be places in rational places. For example, temperatures sensors must be located far from heating or cooling sources, while CO_2 ones should not be placed near windows or where occupants can directly breathe on them [32]. Moreover, it is important to educate occupants to not obstruct, move or hurt the sensors since such perturbations in the data can be difficulty recognised [92].

Outdoor variables are almost always essential. A weather station on the monitoring site is the preferred solution [93] but researchers can also acquire the data from public monitoring stations (e.g., ownership of the Civil Defence) located in the surrounding of the surveyed buildings [94–96].

For detailed thermal evaluations and for the development of accurate behavioural models, the inclusion of outdoor parameters, occurred during the monitoring, is extremely important. However, it should be noted that regulations and BEPS often refer to standard weather data to evaluate building performance. Therefore, the choice of weather data is extremely important in the simulation process, since it should be quite sufficiently meaningful in relation to the case study [97]. In this perspective, many BEPS allow the user to create its own weather file in order to obtain more reliable results.

Figure 6.1 reports an example of monitoring approach in school classrooms. It refers to case study B in Appendix A. The figure shows the plant view of the two rooms with the instruments' position (red dots). Then, the probes and the recorded environmental parameters are presented. In the bottom are displayed a comfort mapping and the mean PMV according to the different zones. These figures have been included to highlight that the thermal behaviour of the room is almost uniform, so the equipment location in the middle of the room is adequately representative of the indoor environment.

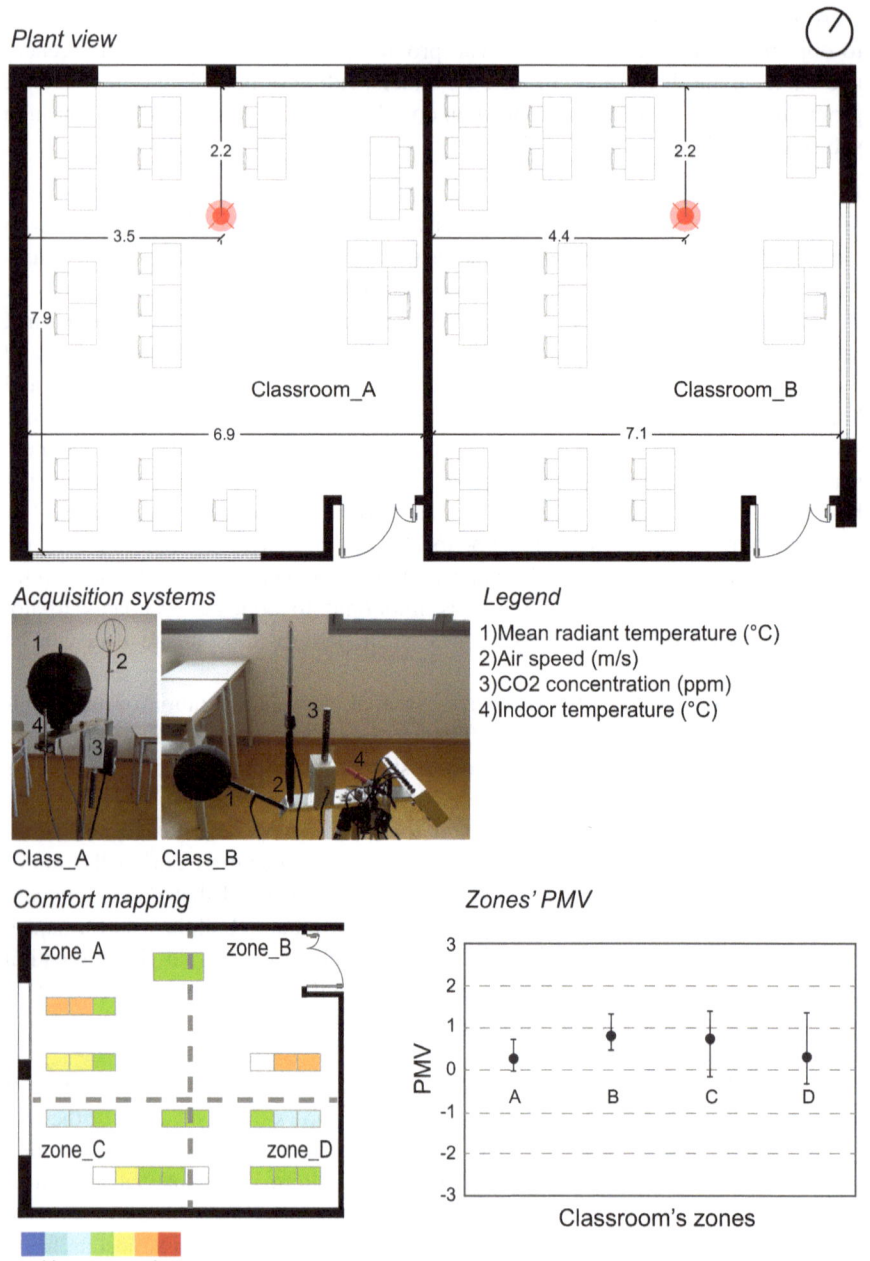

Fig. 6.1 Example of monitoring approach in school classrooms (refer to Appendix A Case study B for further information)

References

1. O'brien W, Gaetani I, Carlucci S et al (2017) On occupant-centric building performance metrics. Build Environ 122:373–385. https://doi.org/10.1016/j.buildenv.2017.06.028
2. Masoso OT, Grobler LJ (2010) The dark side of occupants' behaviour on building energy use. Energy Build 42:173–177. https://doi.org/10.1016/j.enbuild.2009.08.009
3. Webber C, Roberson J, McWhinney M et al (2006) After-hours power status of office equipment in the USA. Energy 31:2823–2838. https://doi.org/10.1016/j.energy.2005.11.007
4. Guan Q, Li C, Guo X, Wang G (2014) Compressive classification of human motion using pyroelectric infrared sensors. Pattern Recognit Lett 49:231–237. https://doi.org/10.1016/j.patrec.2014.07.018
5. Nagy Z, Yong FY, Frei M, Schlueter A (2015) Occupant centered lighting control for comfort and energy efficient building operation. Energy Build 94:100–108. https://doi.org/10.1016/j.enbuild.2015.02.053
6. Shen W, Newsham G, Gunay B (2017) Leveraging existing occupancy-related data for optimal control of commercial office buildings: a review. Adv Eng Informatics. https://doi.org/10.1016/j.aei.2016.12.008
7. Dong B, Andrews B, Lam KP et al (2010) An information technology enabled sustainability test-bed (ITEST) for occupancy detection through an environmental sensing network. Energy Build 42:1038–1046. https://doi.org/10.1016/j.enbuild.2010.01.016
8. Gunay HB, Fuller AF, O'Brien W, Beausoleil-Morrison I (2016) Detecting occupants' presence in office spaces: a case study. eSim 2016
9. Lam KP, Höynck M, Dong B et al (2009) Occupancy detection through an extensive environmental sensor network in an open-plan office building. In: IBPSA Conference, pp 1452–1459
10. Subhan F, Hasbullah H, Rozyyev A, Bakhsh ST (2011) Indoor positioning in Bluetooth networks using fingerprinting and lateration approach. In: 2011 International Conference on Information Science ICISA 2011. https://doi.org/10.1109/ICISA.2011.5772436
11. Li N, Becerik-gerber B (2011) Advanced Engineering Informatics Performance-based evaluation of RFID-based indoor location sensing solutions for the built environment. Adv Eng Informatics 25:535–546. https://doi.org/10.1016/j.aei.2011.02.004
12. Manzoor F, Linton D, Loughlin M (2012) Occupancy monitoring using passive RFID technology for efficient building lighting control. In: RFID Technol, pp 83–88
13. Wang Y, Shao L (2017) Understanding occupancy pattern and improving building energy efficiency through Wi-Fi based indoor positioning. Build Environ 114:106–117. https://doi.org/10.1016/j.buildenv.2016.12.015
14. Richardson I, Thomson M, Infield D (2008) A high-resolution domestic building occupancy model for energy demand simulations. Energy Build 40:1560–1566. https://doi.org/10.1016/j.enbuild.2008.02.006
15. Mahdavi A, Tahmasebi F, Kayalar M (2016) Prediction of plug loads in office buildings: simplified and probabilistic methods. Energy Build 129:322–329. https://doi.org/10.1016/j.enbuild.2016.08.022
16. Gunay HB, O'Brien W, Beausoleil-Morrison I, Gilani S (2016) Development and implementation of an adaptive lighting and blinds control algorithm. Build Environ 113: http://dx.doi.org/10.1016/j.buildenv.2016.08.027
17. Jones RV, Fuertes A, Gregori E, Giretti A (2017) Stochastic behavioural models of occupants' main bedroom window operation for UK residential buildings. Build Environ 118:144–158. https://doi.org/10.1016/j.buildenv.2017.03.033
18. Luo X, Lam KP, Chen Y, Hong T (2017) Performance evaluation of an agent-based occupancy simulation model. Build Environ 115:42–53. https://doi.org/10.1016/j.buildenv.2017.01.015

19. Hailemariam E, Goldstein R, Attar R, Khan A (2011) Real-time occupancy detection using decision trees with multiple sensor types. In: Procs 2011 symposium on simulation for architecture and urban design, pp 141–148
20. Nassif N (2011) CO_2—based demand-controlled ventilation control strategies for multi-zone HVAC systems
21. Chen Z, Soh YC (2016) Comparing occupancy models and data mining approaches for regular occupancy prediction in commercial buildings. J Build Perform Simul 1493:1–9. https://doi.org/10.1080/19401493.2016.1199735
22. Wang D, Arens E, Federspiel C (2003) Opportunities To save energy and improve comfort by using wireless. Icebo 13–18
23. Clements-Croome D (2014) Sustainable intelligent buildings for better health, comfort and well-being
24. Atallah L, Elhelw M, Pansiot J et al (2007) Behaviour profiling with ambient and wearable sensing. In: 4th international workshop on wearable and implantable body sensor networks (BSN 2007), pp 133–138. https://doi.org/10.1007/978-3-540-70994-7_23
25. Zhao Y, Zeiler W, Boxem G, Labeodan T (2015) Virtual occupancy sensors for real-time occupancy information in buildings. Build Environ 93:9–20. https://doi.org/10.1016/j.buildenv.2015.06.019
26. de Bakker C, Aries M, Kort H, Rosemann A (2017) Occupancy-based lighting control in open-plan office spaces: a state-of-the-art review. Build Environ 112:308–321. https://doi.org/10.1016/j.buildenv.2016.11.042
27. Guerra-Santin O, Tweed CA (2015) In-use monitoring of buildings: an overview of data collection methods. Energy Build 93:189–207. https://doi.org/10.1016/j.enbuild.2015.02.042
28. Guerra Santin O, Itard L, Visscher H (2009) The effect of occupancy and building characteristics on energy use for space and water heating in Dutch residential stock. Energy Build 41:1223–1232. https://doi.org/10.1016/j.enbuild.2009.07.002
29. McCarney R, Warner J, Iliffe S et al (2007) The Hawthorne Effect: a randomised, controlled trial. BMC Med Res Methodol 7:30. https://doi.org/10.1186/1471-2288-7-30
30. Gunay HB, O'Brien W, Beausoleil-Morrison I (2013) A critical review of observation studies, modeling, and simulation of adaptive occupant behaviors in offices. Build Environ 70:31–47. https://doi.org/10.1016/j.buildenv.2013.07.020
31. Hong T, Yan D, D'Oca S, Chen C (2016) Ten questions concerning occupant behavior in buildings: the big picture. Build Environ 114:518–530. https://doi.org/10.1016/j.buildenv.2016.12.006
32. Gilani S, O'Brien W (2016) Review of current methods, opportunities, and challenges for in-situ monitoring to support occupant modelling in office spaces. J Build Perform Simul: 1–27. https://doi.org/10.1080/19401493.2016.1255258
33. Haldi F, Robinson D (2008) On the behaviour and adaptation of office occupants. Build Environ 43:2163–2177. https://doi.org/10.1016/j.buildenv.2008.01.003
34. Sadeghi SA, Awalgaonkar NM, Karava P, Bilionis I (2017) A Bayesian modeling approach of human interactions with shading and electric lighting systems in private offices. Energy Build 134:185–201. https://doi.org/10.1016/j.enbuild.2016.10.046
35. Schakib-Ekbatan K, Çakıcı FZ, Schweiker M, Wagner A (2015) Does the occupant behavior match the energy concept of the building?—analysis of a German naturally ventilated office building. Build Environ 84:142–150. https://doi.org/10.1016/j.buildenv.2014.10.018
36. Corry E, Pauwels P, Hu S et al (2015) A performance assessment ontology for the environmental and energy management of buildings. Autom Constr 57:249–259. https://doi.org/10.1016/j.autcon.2015.05.002
37. Reinhart CF, Voss K (2003) Monitoring manual control of electric lighting and blinds. Light Res Technol 35:243–258. https://doi.org/10.1191/1365782803li064oa
38. ASHRAE (2010) ASHRAE Standard 55-2010. Thermal environmental conditions for human occupancy, ASHRAE Atlanta, GA. doi:ISSN 1041-2336

39. Ahmad MW, Mourshed M, Mundow D et al (2016) Building energy metering and environmental monitoring—a state-of-the-art review and directions for future research. Energy Build 120:85–102. https://doi.org/10.1016/j.enbuild.2016.03.059
40. Haldi F, Robinson D (2009) Interactions with window openings by office occupants. Build Environ 44:2378–2395. https://doi.org/10.1016/j.buildenv.2009.03.025
41. Rijal HB, Tuohy PG, Humphreys M et al (2007) Using results from field surveys to predict the effect of open windows on thermal comfort and energy use in buildings. Energy Build 39:823–836. https://doi.org/10.1016/j.enbuild.2007.02.003
42. D'Oca S, Hong T (2014) A data-mining approach to discover patterns of window opening and closing behavior in offices. Build Environ 82:726–739. https://doi.org/10.1016/j.buildenv.2014.10.021
43. Hargreaves T, Hauxwell- R, Stankovic L et al (2015) Smart homes, control and energy management: how do smart home technologies influence control over energy use and domestic life ? In: European Council for an Energy Efficient Economy 2015 Summer Study Energy Efficient, pp 1021–1032
44. Habara H, Yasue R, Shimoda Y (2013) Survey on the occupant behavior relating to window and air conditioner operation in the residential buildings. In: 13th Conference International Building Performance Simulation Association, pp 2007–2013
45. Bourikas L, Costanza E, Gauthier S et al (2016) Camera-based window-opening estimation in a naturally ventilated office. Build Res Inf 1–16. https://doi.org/10.1080/09613218.2016.1245951
46. Yan D, O'brien W, Hong T et al (2015) Occupant behavior modeling for building performance simulation: current state and future challenges. Energy Build 107:264–278. https://doi.org/10.1016/j.enbuild.2015.08.032
47. Vogit J (2015) Angular positioning of a door or window using a MEMS accelerometer and a magnetometer. Lund University
48. Zhang Y, Barrett P (2012) Factors influencing the occupants' window opening behaviour in a naturally ventilated office building. Build Environ 50:125–134. https://doi.org/10.1016/j.buildenv.2011.10.018
49. Inkarojrit V, Paliaga G (2004) Indoor climatic influences on the operation of windows in a naturally ventilated building. In: Proceedings of 21th International Conference on Passive and Low Energy Architecture, Netherlands, pp 427–431
50. Raja IA, Nicol JF, McCartney KJ, Humphreys MA (2001) Thermal comfort: use of controls in naturally ventilated buildings. Energy Build 33:235–244. https://doi.org/10.1016/S0378-7788(00)00087-6
51. Correia da Silva P, Leal V, Andersen M (2013) Occupants interaction with electric lighting and shading systems in real single-occupied offices: results from a monitoring campaign. Build Environ 64:152–168. https://doi.org/10.1016/j.buildenv.2013.03.015
52. Yao J (2014) Determining the energy performance of manually controlled solar shades: a stochastic model based co-simulation analysis. Appl Energy 127:64–80. https://doi.org/10.1016/j.apenergy.2014.04.046
53. O'Brien W, Kapsis K, Athienitis AK (2013) Manually-operated window shade patterns in office buildings: a critical review. Build Environ 60:319–338. https://doi.org/10.1016/j.buildenv.2012.10.003
54. Haldi F, Robinson D (2010) Adaptive actions on shading devices in response to local visual stimuli. J Build Perform Simul 3:135–153
55. Sutter Y, Dumortier D, Fontoynont M (2006) The use of shading systems in VDU task offices: a pilot study. Energy Build 38:780–789. https://doi.org/10.1016/j.enbuild.2006.03.010
56. Foster M, Oreszczyn T (2001) Occupant control of passive systems: the use of Venetian blinds. Build Environ 36:149–155. https://doi.org/10.1016/S0360-1323(99)00074-8
57. Rea MS (1984) Window blind occlusion: a pilot study. Build Environ 19:133–137. https://doi.org/10.1016/0360-1323(84)90038-6

58. Rubin AI, Collins BL, Tibbott RL (1978) Window blinds as a potential energy saver-a case study. Natiolal Bureau of Standards, Washington DC
59. Zhang Y, Barrett P (2012) Factors influencing occupants' blind-control behaviour in a naturally ventilated office building. Build Environ 54:137–147. https://doi.org/10.1016/j. buildenv.2012.02.016
60. Moore T, Carter DJ, Slater A (2003) Long-term patterns of use of occupant controlled office lighting. Light Res Technol 35:43–59. https://doi.org/10.1191/1477153503li061oa
61. Yun GY, Kim H, Kim JT (2012) Effects of occupancy and lighting use patterns on lighting energy consumption. Energy Build 46:152–158. https://doi.org/10.1016/j.enbuild.2011.10.034
62. Hunt DRG (1979) The use of artificial lighting in relation to daylight levels and occupancy. Build Environ 14:21–33. https://doi.org/10.1016/0360-1323(79)90025-8
63. Weilil JS, Gladhart PM (1990) Occupant behavior and successful energy conservation: findings and implications of behavioral monitoring, pp 171–180
64. Woods J (2006) Fiddling with Thermostats: energy Implications of Heating and Cooling Set Point Behavior. 2006 ACEEE Summer Study Build, pp 278–287
65. Zhou X, Yan D, Feng X et al (2016) Influence of household air-conditioning use modes on the energy performance of residential district cooling systems. Build Simul 9:429–441. https://doi.org/10.1007/s12273-016-0280-9
66. Ren X, Yan D, Wang C (2014) Air-conditioning usage conditional probability model for residential buildings. Build Environ 81:172–182. http://dx.doi.org/10.1016/j.buildenv.2014. 06.022
67. Tanimoto J, Hagishima A (2005) State transition probability for the Markov Model dealing with on/off cooling schedule in dwellings. Energy Build 37:181–187. https://doi.org/10.1016/ j.enbuild.2004.02.002
68. Bae C, Chun C (2009) Research on seasonal indoor thermal environment and residents' control behavior of cooling and heating systems in Korea. Build Environ 44:2300–2307. https://doi.org/10.1016/j.buildenv.2009.04.003
69. Karjalainen S (2009) Thermal comfort and use of thermostats in Finnish homes and offices. Build Environ 44:1237–1245. https://doi.org/10.1016/j.buildenv.2008.09.002
70. Peffer T, Pritoni M, Meier A et al (2011) How people use thermostats in homes: a review. Build Environ 46:2529–2541. https://doi.org/10.1016/j.buildenv.2011.06.002
71. Nicol JF (2001) Characterising occupant behavior in buildings: towards a stochastic model of occupant use of windows, lights, blinds heaters and fans. In: 7th International IBPSA Conference, pp 1073–1078
72. Burak Gunay H, O'Brien W, Beausoleil-Morrison I, Perna A (2014) On the behavioral effects of residential electricity submetering in a heating season. Build Environ 81:396–403. https:// doi.org/10.1016/j.buildenv.2014.07.020
73. Teli D, Jentsch MF, James PAB (2012) Naturally ventilated classrooms: an assessment of existing comfort models for predicting the thermal sensation and preference of primary school children. Energy Build 53:166–182. https://doi.org/10.1016/j.enbuild.2012.06.022
74. Gonçalves H, Ocneanu A, Bergés M, Fan RH (2011) Unsupervised disaggregation of appliances using aggregated consumption data. Environ Eng
75. Schiavon S, Lee KH (2013) Dynamic predictive clothing insulation models based on outdoor air and indoor operative temperatures. Build Environ 59:250–260. https://doi.org/10.1016/j. buildenv.2012.08.024
76. Morgan C, de Dear R (2003) Weather, clothing and thermal adaptation to indoor climate. Clim Res 24:267–284. https://doi.org/10.3354/cr024267
77. ISO (2007) 9920:2007, Ergonomics of the thermal environment—estimation of thermal insulation and water vapour resistance of a clothing ensemble
78. ISO (2006) 7730:2006 Ergonomics of the thermal environment—Analytical determination and interpretation of the thermal comfort using calculation of the PMV and PPD indices and local thermal comfort criteria
79. CEN (2005) UNI EN ISO 8996 Ergonomics of the thermal environment—determination of metabolic rate. Brussels

80. Revel GM, Arnesano M, Pietroni F (2014) A low-cost sensor for real-time monitoring of indoor thermal comfort for ambient assisted living. In: Ambient Assistance Living, pp 3–12
81. Hasan MH, Alsaleem F, Rafaie M (2016) Sensitivity study for the PMV thermal comfort model and the use of wearable devices biometric data for metabolic rate estimation. Build Environ 110:173–183. https://doi.org/10.1016/j.buildenv.2016.10.007
82. Luo M, Ji W, Cao B et al (2016) Indoor climate and thermal physiological adaptation: evidences from migrants with different cold indoor exposures. Build Environ 98:30–38. https://doi.org/10.1016/j.buildenv.2015.12.015
83. Van Den Wymelenberg K, Inanici M (2009) A study of luminance distribution patterns and occupant preference in Daylit Offices. Energy 22–24. https://doi.org/10.1582/LEUKOS.2010.07.02003
84. Gao J, Wargocki P, Wang Y (2014) Ventilation system type, classroom environmental quality and pupils' perceptions and symptoms. Build Environ 75:46–57. https://doi.org/10.1016/j.buildenv.2014.01.015
85. Humphreys MA, Rijal HB, Nicol JF (2013) Updating the adaptive relation between climate and comfort indoors; new insights and an extended database. Build Environ 63:40–55. https://doi.org/10.1016/j.buildenv.2013.01.024
86. Langevin J, Gurian PL, Wen J (2015) Tracking the human-building interaction: a longitudinal field study of occupant behavior in air-conditioned offices. J Environ Psychol 42:94–115. https://doi.org/10.1016/j.jenvp.2015.01.007
87. Stazi F, Naspi F, Ulpiani G, Perna C Di (2017) Indoor air quality and thermal comfort optimization in classrooms developing an automatic system for windows opening and closing. Energy Build 139:732–746. https://doi.org/10.1016/j.enbuild.2017.01.017
88. Meerbeek B, te Kulve M, Gritti T et al (2014) Building automation and perceived control: a field study on motorized exterior blinds in Dutch offices. Build Environ 79:66–77. https://doi.org/10.1016/j.buildenv.2014.04.023
89. Stazi F, Naspi F, D'Orazio M (2017) A literature review on driving factors and contextual events influencing occupants' behaviours in buildings. Build Environ 118:40–66. https://doi.org/10.1016/j.buildenv.2017.03.021
90. Fabi V, Andersen RV, Corgnati S, Olesen BW (2012) Occupants' window opening behaviour: a literature review of factors influencing occupant behaviour and models. Build Environ 58:188–198. https://doi.org/10.1016/j.buildenv.2012.07.009
91. ISO (2001) 7726:2001 Ergonomics of the thermal environment—instruments for measuring physical quantities
92. O'Brien W, Gunay HB (2014) The contextual factors contributing to occupants' adaptive comfort behaviors in offices—a review and proposed modeling framework. Build Environ 77:77–88. https://doi.org/10.1016/j.buildenv.2014.03.024
93. Pisello AL, Castaldo VL, Piselli C et al (2016) How peers' personal attitudes affect indoor microclimate and energy need in an institutional building: results from a continuous monitoring campaign in summer and winter conditions. Energy Build 126:485–497. https://doi.org/10.1016/j.enbuild.2016.05.053
94. Stazi F, Naspi F, D'Orazio M (2017) Modelling window status in school classrooms. Results from a case study in Italy. Build Environ 111:24–32. https://doi.org/10.1016/j.buildenv.2016.10.013
95. Yun GY, Steemers K (2008) Time-dependent occupant behaviour models of window control in summer. Build Environ 43:1471–1482. https://doi.org/10.1016/j.buildenv.2007.08.001
96. D'Oca S, Fabi V, Corgnati SP, Andersen RK (2014) Effect of thermostat and window opening occupant behavior models on energy use in homes. Build Simul 7:683–694. https://doi.org/10.1007/s12273-014-0191-6
97. Roetzel A, Tsangrassoulis A, Dietrich U, Busching S (2010) A review of occupant control on natural ventilation. Renew Sustain Energy Rev 14:1001–1013. https://doi.org/10.1016/j.rser.2009.11.005

Chapter 7
Modelling, Implementation and Validation Approaches

Abstract Behavioural models are fundamental tools to predict users' behaviours and evaluate buildings' performance both during the design and operation phases. Occupancy patterns and human-building interaction have been forecasted according to several approaches, which lead to different levels of accuracy and reliability. Such models improve their usefulness when implemented in Building Energy Performance Simulation (BEPS) that, coupling specific behavioural tools with environmental simulations, can enhance the results and bridge the performance gap.

7.1 Introduction

Many approaches have been proposed for the development of occupancy and behavioural models aimed at predicting users' presence and actions inside buildings. Early deterministic approaches, based on fixed and static rules have been increasingly replaced by more accurate and precise models. A further step has been obtained defining different users' types or behavioural lifestyles, which cluster occupants' energy-related attitudes. Stochastic models represent the variability of people's behaviours even more. To better reflect users' real actions, stochastic models are usually developed using data acquired during experimental campaigns. However, the lacking of a standardised method led the researchers to propose and test a great variety of approaches. Also the models' implementation in BEPS and their validation processes have been performed in numerous ways. Offering a complete overview of these topics, the current Chapter aims at presenting the main strategies to model users' behaviours (from Sects. 7.2–7.4) and occupancy patterns (Sect. 7.5), to embed the models in simulators (Sect. 7.6) and to validate them (Sect. 7.7).

© The Author(s) 2018   63
F. Stazi and F. Naspi, *Impact of Occupants' Behaviour on Zero-Energy Buildings*,
SpringerBriefs in Energy, https://doi.org/10.1007/978-3-319-71867-5_7

7.2 Behavioural Modelling Approaches

The modelling of occupants' behaviours in buildings can be performed following three main strategies. These approaches are sketched in Fig. 7.1 and discussed below, starting from the easiest.

The deterministic approach

Early, the behavioural models were developed according to pre-determined laws. For example, the light-switching model proposed by Hunt [1] defined the lights status according to different times of the day. At the first entrance, the lights were switched on, during the occupied periods were maintained unchanged and switching off behaviours were only connected to departures.

Such models, remaining anyway deterministic, can be improved with the combination of *static* and environmentally-driven *rules*. Usually, a fixed threshold of one or more environmental triggers is included in the model framework. For example, the above-mentioned model [1] was enhanced including a rule based on the work-plane illuminance level (i.e. the switch-on probability at arrival and after lunch is calculated only for illuminance levels below of 150 lx) [2]. Similar methods are currently available in almost all BEPS, especially to assess window opening and closing. For example, windows are considered open when the indoor temperature is higher than the outdoor one and higher than a fixed limit (usually between 24 and 26 °C).

A further step concerns the adoption of *decision-trees methods*. Using this approach, the generated model has flowchart-like tree form, aiding the users to extract suitable information and to perform classifications and predictions with

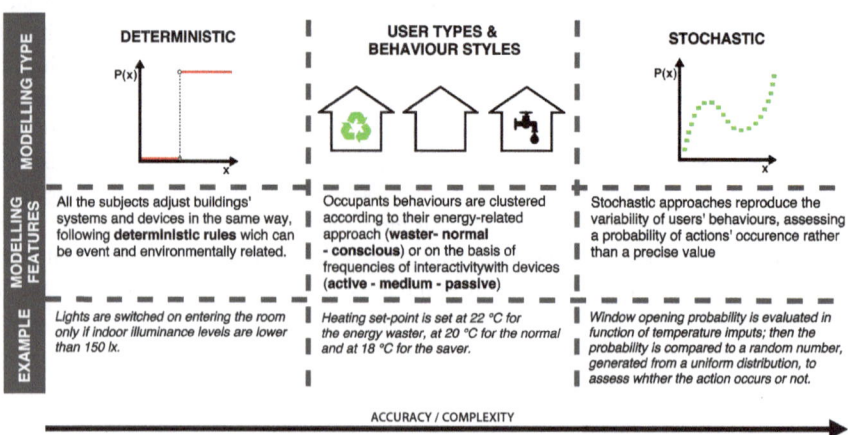

Fig. 7.1 Sketch of the most diffused approaches to model people's behaviour in buildings

minimal computation efforts [3]. Accurate predictions of residential building energy performance indexes have been estimated using this method.

User types and behaviour styles
Deterministic approaches start from the assumption that all the users behave in the same way and uniformly react to the stimuli. Even if static rules are the easiest solutions both in terms of development and implementation, numerous studies [4, 5] highlighted that people have different preferences, needs and reactions to external triggers. To include such occupants' diversities, Behavioural Patterns and Users' Profiles have been proposed. These methods cluster the occupants according to energy-related macro categories. In fact, different background habits, lifestyles and comfort preferences influence behaviours and, consequently, the overall energy use.

Behaviour styles [6–8] are related to the users' approach regarding energy (e.g. waster, average, saver), while user types [9–11] are connected to occupants' variability on buildings' controls (i.e. active, medium and passive). The first type of approach is mainly adopted to investigate how much different environmental settings (e.g. thermostat set-point) affect the energy use. The second one, while having similar aims, clusters the occupants according to the frequency with which they adjust the surrounding. The category definition has been performed both following personal responses from questionnaires [12, 13], using objective data (e.g. number of actions recorded by the sensors) [10, 14, 15], as well as adopting numerical indicators [9].

Simulation studies [6], performed in the people clustering perspective, highlighted that when the human-building interaction is great the potential energy savings are significantly affected by occupants' behaviour (e.g. energy spenders consume twice as savers). In addition, behavioural patterns and user profiles (i.e. users grouped according to their characteristics) could be linked, in the perspective of both improve simulated predictions and formulate energy-saving policy for specific social groups [16].

Stochastic approaches
Despite the above-mentioned approaches provide more reliable results in comparison to deterministic ones, they still categorise occupants and behaviours according to static classes, providing usually only boundary conditions. For these reasons, stochastic approaches are preferred to reproduce users' random nature. Probabilistic models can be adopted using three main forms: Bernoulli processes, discrete-time Markov chains and survival analyses.

The *Bernoulli models* predict the state of a building system or device in function of one or more predictor variables [17]. The main feature of these processes is the memoryless hypothesis, according to which the decisions taken at each time step are independent of the previous and so, there is no direct influence between different moments. They have been widely adopted for their computational efficiency and for the need of quite few information. They are suitable for energy modelling at a large scale (e.g. building level) but they are poor at individual scale [18]. In fact, since occupants have a different degree of interaction with opposite actions (e.g. close the blinds to avoid glare and then wait days until reopen them), such models tend to overestimate users' interactivity [19].

The second modelling approach regards the adoption of discrete-time *Markov chains*. On the contrary of Bernoulli process, Markov chains follow a memory approach for which the action at each time step is influenced from (at least) the previous one. In fact, Markov processes can be divided between homogeneous [20] or inhomogeneous chains [21]. In the former case, the transition probability depends only on the system state at the previous time step, while in the second case it is influenced by the history of the system, meaning that every single state is a consequence of the entire process. This memory property offers a better representation of individuals' actions, generating realistic patterns [17]. Issues of this approach are related to the dependency on fixed time-steps and to the computational effort that scales linearly with the number of modelled occupants [22].

Survival analyses are the third behaviour modelling approach. Usually, this method is adopted for assessing the time duration until an event occurs, estimating how long a building or a device remains unchanged. For example, they can be used to predict the duration of an occupied period upon arrival [23] or the duration of a window opening [9]. Survival analyses could be particularly useful when the driving factors have a low and weak influence on occupants; in fact, since this is a continuous time approach, periods of elapsed time should be defined before an action occurs [24].

When one of the above-mentioned stochastic approaches is selected to predict users' behaviours, some considerations on the simulation process are needed. In fact, it should be noted that, at each time-step, the output is not a binary value which clearly defines whether the condition is true or false but it is a probability, ranged between 0 and 100%. To assess the occurrence of an event (e.g. window opening) or the occupancy status (i.e. user presence/absence), the probabilities should be compared to a series of numbers randomly generated from a uniform distribution [25]. If the probability is higher than the random number, the output is true (e.g. the window turns the status from close to open), otherwise, it is false. In this perspective of obtaining reliable stochastic results, a certain number of simulations are necessary [26]. In fact, after the run of N simulations, the performance indicators (e.g. energy consumptions) take the form of probability distributions instead of single values [14]. Figure 7.2 shows how the stochastic approach works, highlighting that the behaviours at each time step change due to the probabilistic nature of the functions and that also the final result has a stochastic form.

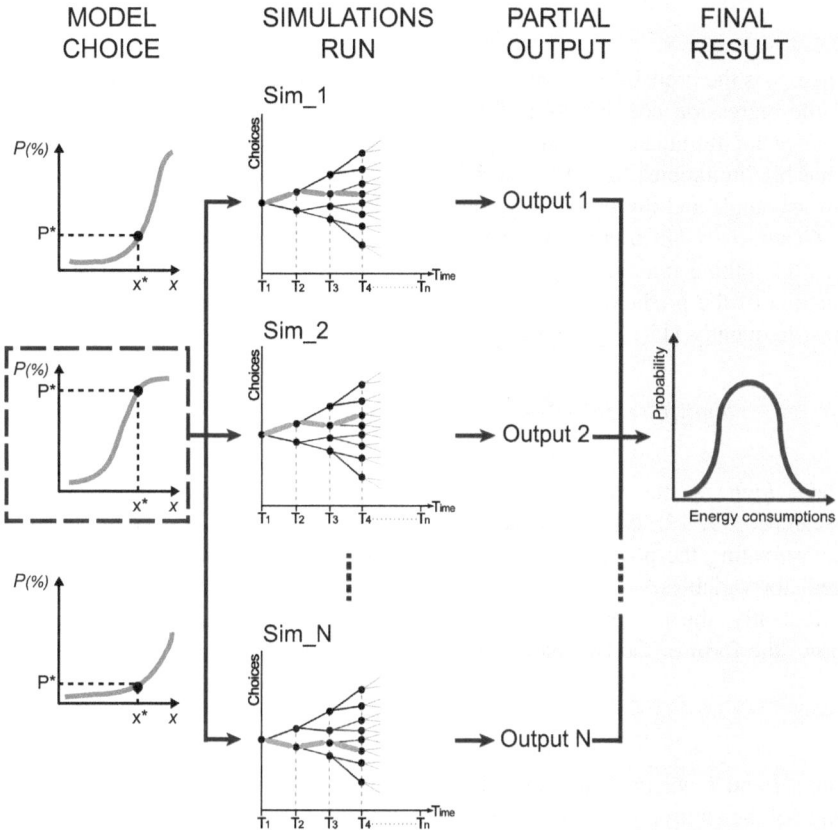

Fig. 7.2 The adoption of a stochastic approach requires several simulations of the same environment since each simulation can produce different results. As a consequence, the main result would be a distribution of energy consumptions instead of a single value

7.3   The Mathematical Form

The choice of a stochastic approach (e.g. Bernoulli process or Markov chains) is followed by the selection of the mathematical form. The equation should fit the experimental data as well as possible and, in parallel, it must not present physical impossibilities for the parameters ranges (e.g. probabilities greater than 1). Usually, the equation mathematical form is determined by evaluating the fit for several generic formulations. Some researchers [2, 27] suggested the adoption of *linear models* to estimate the actions' probability as a function of one or more predictor variables. The linear relationship between the response and the predictors is expressed by the following equation:

$$p_i = \beta_0 + \beta_1 x_{1,i} + \beta_2 x_{2,i} + \cdots + \beta_n x_{n,i} \tag{7.1}$$

where p_i is the probability of action occurrence (e.g. light switching), β is the vector of the regression coefficients and x is the vector of the predictor variables (e.g. work-plane illuminance). Even if this approach is extremely simple and immediate, it has big limitations. In fact, it poorly predicts the observations at the upper and the lower bounds and the probability can be greater than one [28].

Generalised linear models have been introduced to overcome these issues. In fact, the linking function (e.g. logit and probit) of the response variable is a linear function of the predictor variables. Logistic regression models, the forms adopted most frequently [12, 24], are expressed as follows:

$$logit(p_i) = ln\left(\frac{p_i}{1 - p_i}\right) = \beta_0 + \beta_1 x_{1,i} + \beta_2 x_{2,i} + \cdots + \beta_n x_{n,i} \tag{7.2}$$

where *logit* is the linking function for logistic regression. Currently, many researchers [9, 12, 29] have evaluated logistic regression models appropriateness for estimating the probability of an adaptive behaviour triggered by one or more predictor variables.

Recently, the use of Weibull distribution functions is increased. The Eq. 7.3 shows the form of the two parameters Weibull distribution:

$$p = 1 - e^{-\left(\frac{x}{l}\right)^k} \tag{7.3}$$

where l and k are the scale and shape parameters. The first represents the linear effect of the environmental stimulus and the second its power effect. The use of this form has been suggested by Haldi and Robinson [9] to predict window opening duration and by Mahdavi et al. [30] for plug loads in offices.

Three parameters distributions have been proposed too. This approach has been used to develop a generalised probabilistic formula which predicts users' behaviours on building devices [31–33], introducing a conditional probability.

$$p = \begin{cases} 1 - e^{-\left(\frac{x-u}{l}\right)^k}, & x \geq u \\ 0, & x < u \end{cases} \tag{7.4}$$

Equation 7.4 shows that the third parameter u is the threshold for the action occurrence, typical of occupants' physical response to the trigger x. The three parameters are experimentally defined and they identify how much the sample is influenced by a peculiar stimulus. The formulas were tested to predict light, window and the air-conditioning use according to both environmental and event triggers. Figure 7.3 shows a graphical representation of the three above-mentioned formulas (i.e. linear, logistic and Weibull).

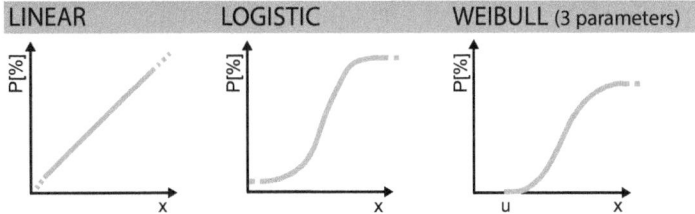

Fig. 7.3 Graphical representation of the main formulas adopted to develop behavioural models

Other mathematical forms are less diffused. Quadratic equations [34] and probit regressions [35] have been proposed to predict window opening, while sigmoid functions were adopted to evaluate shadings and lights adjustments [36].

7.4   Advanced Stochastic Modelling: Agent-Based Models

To improve stochastic approaches reliability, probabilistic rules can be included in advanced modelling, as Agent Based Modelling (ABM). This approach, which can be seen as an extension of Markovian models, can be useful to catch the diversity of modelled occupants, to investigate occupancy patterns and several actions simultaneously.

ABM allows moving from the group to the individual level since each occupant is represented as an "agent" with memory that interacts with the other agents and with the surrounding, following pre-determined rules [37]. These features permit to reflect with more accuracy people's dynamic and variable behaviours [38, 39].

The decision rules have been developed according to the PMV comfort model [39, 40] or following the Perceptual Control Theory [41], which is based on the assumption that thermal comfort is restored taking the most direct action. A further decisional approach concerns the Belief-Desire-Intention (BDI) model [42], that target at reproducing a rational human decision-making process on the bases of personal values and society's norms [18].

It is evident that a standardised method is still lacking and so, different results can be difficultly compared. In addition to the peculiar comfort approaches, others limitations and restrictions are related to ABMs. When ABMs are developed using real data (i.e. empirical models), better results are provided in comparison to abstract ones. However, empirical models need big data sets and very precise rules to be reliable. The main consequence is the increasing of model's complexity, which can also lead to difficulties in initialisations and lacks in clearness and completeness [39]. Moreover, it has been assessed that a nonlinear relationship correlates the complexity of the model behaviour and the complicatedness of model structure. This implies that little improvements of the former cause very big increase of the latter [43]. Therefore, the choice of adopting the ABM approach should be coherent with the target of the research and with the expected results [44].

7.5 Approaches to Model Occupancy Patterns

The study and analysis of occupancy patterns have been the main target of several researchers for twofold reasons. The first is that people's presence influences the use of building systems (e.g. HVAC), devices (e.g. windows and shadings), plug loads (e.g. PC) and, as a consequence, overall energy consumptions [45]. The second reason is linked to the optimisation of building performance since a correct knowledge of users' presence and absence can avoid energy waste in unoccupied spaces.

Currently, the inclusion of occupancy patterns in the building management occurs through static and pre-definite schedules. However, the adoption of occupancy sensors can improve the building energy use, allowing especially to adjust lighting and ventilation instantaneously [46]. In the perspective of predicting building performance and developing behavioural models, occupants' presence is a crucial data since it is the main precondition for actions' occurrence. Although many researchers tried to reproduce users' presence, overall methods and findings are still lacking. To overcome such limitations, different modelling approaches have been provided. In the following, the most significant models are briefly discussed in order of increasing complexity.

Stochastic models are usually preferred to reproduce the variability of people's location [47]. Among them, Markovian models, which predict the transition probability from a state to another, seem to be rather diffuse [48]. Both homogeneous and inhomogeneous chains have been adopted to simulate the occupants' movement [20] and presence/vacancy [49], respectively. The second approach, considering all the previous time-steps, seem to have better performance. In fact, inhomogeneous Markov chains have also been used to predict plug loads patterns in office buildings generating non-repeating daily occupancy profiles [30].

Other authors adopted non-probabilistic models but their diffusion is limited. In [50] the predictions of a new simple non-probabilistic model, which generates binary occupancy daily profiles, are compared to two stochastic models from the literature. Results highlighted that both the non-probabilistic and the stochastic models have low rates of accuracy but the first shows better predictive performances.

Data-mining techniques and analyses of big data stream have been recognised as powerful methods to identify and reproduce correlations among different occupancy states [51, 52]. When big data-sets are available, data-driven modelling can be a valid alternative. Data-driven models, requiring prior monitored data, use a given set of laws of a specific case to predict its future behaviour [53]. In the perspective of enhance the reliability of estimated occupancy patterns, data-driven models have been combined with an occupancy model [54]. Similarly, a toolkit which contains 22 Matlab functions has been proposed using a data-driven approach to conduce statistical analyses on both occupancy and behavioural patterns [17].

The stochasticity of occupants' presence has been reproduced also using ABMs. For example, a model to evaluate the location of each occupant at each time step on the basis of a set of rules has been proposed, also including a graphical modelling framework to create a simple occupancy model in a multi-zone building [55]. Recently, also software models and free web applications [56, 57] have been made available to generate occupancy profiles, simulating users' presence and movement.

7.6 Implementing Behavioural Models in BEPS

BEPS are widely used to evaluate building performances, especially during the design phase. However, results from simulations are often very far from real ones [58]. One of the main identified reasons for this gap is the incorrect representations of users' stochastic behaviours [18]. In fact, the behavioural models' implementation in BEPS is still uncommon, despite their promising predicting capabilities have been widely recognised.

Almost all the existing simulation programs adopted a *deterministic* and *static approach* to represent users' actions and occupancy patterns [5, 59]. Generally, the user can set standard schedules and, sometimes, slightly modify the deterministic rules that mime occupants' actions (e.g. define the threshold for window opening). The main advantage of adopting deterministic rules is their easiness in use and interpretation. In fact, also non-expert modellers can quickly modify them to represent lighting and plug load usage, interactions with natural ventilation and HVAC systems, and occupancy patterns. Moreover, the replication of the behavioural phenomena is supported by many experimental studies on responses of human physiology, as the huge campaigns performed to develop both the PMV and the adaptive comfort models [40, 60]. However, major drawbacks of this modelling approach are steady-periodicity and weakness in capturing and reproducing human stochasticity, causing big discrepancies between real and simulated results.

Improvements can be reached by *modifying* or *overwriting* the program *source code* and, so implementing a developed behavioural model [61]. This technique usually requires a high expertise of the program and only few software allow such customisation (e.g. EnergyPlus, ESP-r, TRNSYS). An example of customisation is provided by the energy management system (EMS) in EnergyPlus. EMS is a high-level control method, which offers the possibility to tune the same controls available through EMS in real buildings [62].

Many authors [53, 63] affirmed that discrepancies between BEPS-predicted and actual metered building energy use can be resolved through the adoption of a *co-simulation* approach.

Using the co-simulation, it would be possible to run simulations in which the behavioural model and the environmental simulation exchange data in real time. Even if the implementation of this method is not immediate and requires expertise, during recent years many efforts have been made to develop and implement it in BEPS, aiming at reaching more reliable simulation results than in the past. Moreover, encouraging the interoperability between behavioural models and BEPS, co-simulation provides high levels of flexibility to the users and potential use in BIM [33].

Researchers which proposed the co-simulation approach usually couple EnergyPlus with several software utilities: the Functional Mock-up Unit (FMU) [33, 64], the Building Controls Virtual Test Bed (BCVTB) [65, 66], CHAMPS-Multizone [67] and Brahms environment [68].

In the perspective of developing a standardised method, the international project Annex 66 [69] proposed an ontology to represent occupants' behaviours in buildings. The DNAs framework [70] consists of the Drivers of users' behaviours, the Needs of the occupants, the Actions people take to restore comfort conditions and the adjusted building systems. The framework has been also represented using an XML (eXtensible Markup Language) schema, directed to implementation in BEPS or in Functional Mock-up Units [71]. The embedding of the XML schema inside FMU, using a functional mock-up interface (FMI) [33], was aimed at providing the co-simulation both with EnergyPlus and ESP-r. This functionality is likely to be reasonably intuitive and well supported. The occupant behaviour FMU has a good portability since it is distributed with a library of existing occupant behaviour models. Further steps have been made publicly proposing a new library with 52 behavioural models [72]. Collected from the literature and formatted according to the XML schema, they target at providing various valid examples of human-building interaction.

BEPS improved with behavioural models can also be used during buildings' operational phase to detect faults, tuning operations and suggest measures to improve building performances [73].

7.7   Behavioural Models Evaluation and Validation

The behavioural models validation process is still few addressed since there is no a whole accepted method. This lack is also a consequence of the missing standardization of behavioural models development.

The model evaluation process should preferably use observational data not already adopted in the model development process. The report which explains the model evaluation should define all the model specifics and the limitations in order to allow further researcher to know the correct way to perform it [18].

The most accurate validation process concerns the use of two (or more) datasets, obtained from independent (but similar) contexts. In fact, in this case, it is possible to develop two different models and test each of them on the other sample [74, 75]. Thus, the model predictability can be checked on another building with different occupants and behaviours. This type of verification is very limited in the literature, since such independent datasets are difficultly available [24]. For example, the Smart Controls and Thermal Comfort (SCAT) dataset have been used to test other models [74, 76].

When only one dataset is available, the three main types of validation are commonly adopted [77]:

(1) *Validation set approach.* The dataset is split in two independent set: the training set and the validation one. The model is developed using the former and is tested using the latter. The test error is frequently estimated through the MSE (mean squared error). The sampling method is crucial for the reliability of the results. Since the two sets can be variously defined, the method presents a great uncertainty. Moreover, a large amount of data is required since only half of the sample is used to develop the model. While with insufficient data, the model can be not accurate.

(2) *Leave-one-out cross validation* (LOOCV). The dataset is (usually) divided in ten sub-sets: nine observations are used to develop the model (e.g. $\{(x_2,y_2),\ldots,(x_n,y_n)\}$) and one is left out for the validation [e.g. (x_1,y_1)]. The predicted value y_1^* (e.g. window opening probability) is calculated using x_1 (e.g. a value of indoor temperature), since it was not used in the model estimation, and the MSE is calculated [i.e. $(y_1 - y_1^*)$]. The process is repeated the same number of the sub-sets (in the case, ten times), to use the entire sub sample both as training and validation set. At the end n MSE are calculated, and so, the overall value for the cross validation (CV) is obtained (i.e. $CV_{(n)} = \frac{1}{n}\sum_{i=1}^{n} MSEi$). This method has less bias in comparison to the validation set approach and so, the test error is less overestimated. The results have no variability for the same initial dataset since the entire sample is adopted. The main disadvantage of the method is the intense computational work.

(3) *K-fold cross validation.* The sample of the observations is casually divided in *k* folders of about the same dimension. The first folder is considered as the validation set and the model is developed using the remaining *k*-folders. After this phase, the MSE is calculated. The entire procedure is then repeated *k* times. At the end, *k* MSE are calculated and the CV is obtained too (i.e. $CV_{(k)} = \frac{1}{k}\sum_{i=1}^{n} MSEi$). This method simplifies the previous one since k<<n.

Cross validation is not frequently adopted in the literature since it is computationally intense. The LOOCV has been used to validate a model to predict the clothing level [74], while the *k*-fold approach to test a window opening one [75].

In general, researchers usually validate the proposed models only comparing the generated or simulated data to the recorded ones [78]. Such technique is the quickest and easiest; it has been used to validate both windows [9, 79] and ABMs [43], comprehensive of many different actions.

References

1. Hunt DRG (1980) Predicting artificial lighting use a method based upon observed patterns of behavior. Light Res Technol 12:7–14. https://doi.org/10.1017/CBO9781107415324.004
2. Newsham GR (1994) Manual control of window blinds and electric lighting: implications for comfort and energy consumption. Indoor Environ 6:135–144. https://doi.org/10.1177/1420326X9400300307

3. Yu Z, Haghighat F, Fung BCM, Yoshino H (2010) A decision tree method for building energy demand modeling. Energy Build 42:1637–1646. https://doi.org/10.1016/j.enbuild. 2010.04.006

4. Andersen RV, Toftum J, Andersen KK, Olesen BW (2009) Survey of occupant behaviour and control of indoor environment in Danish dwellings. Energy Build 41:11–16. https://doi.org/ 10.1016/j.enbuild.2008.07.004

5. Hong T, Yan D, D'Oca S, Chen C (2016) Ten questions concerning occupant behavior in buildings: the big picture. Build Environ 114:518–530. https://doi.org/10.1016/j.buildenv. 2016.12.006

6. Sun K, Hong T (2017) A framework for quantifying the impact of occupant behavior on energy savings of energy conservation measures. Energy Build 146:383–396. https://doi.org/ 10.1016/j.enbuild.2017.04.065

7. Hong T, Lin HW (2013) Occupant behavior: impact on energy use of private offices. Lawrence Berkeley Natl. Lab, Berkeley, CA

8. Karjalainen S (2016) Should we design buildings that are less sensitive to occupant behaviour? A simulation study of effects of behaviour and design on office energy consumption. Energy Effic 9:1257–1270. https://doi.org/10.1007/s12053-015-9422-7

9. Haldi F, Robinson D (2009) Interactions with window openings by office occupants. Build Environ 44:2378–2395. https://doi.org/10.1016/j.buildenv.2009.03.025

10. Yun GY, Tuohy P, Steemers K (2009) Thermal performance of a naturally ventilated building using a combined algorithm of probabilistic occupant behaviour and deterministic heat and mass balance models. Energy Build 41:489–499. https://doi.org/10.1016/j.enbuild.2008. 11.013

11. Reinhart CF (2004) Lightswitch-2002: a model for manual and automated control of electric lighting and blinds. Sol Energy 77:15–28. https://doi.org/10.1016/j.solener.2004.04.003

12. Rijal HB, Tuohy PG, Humphreys M et al (2007) Using results from field surveys to predict the effect of open windows on thermal comfort and energy use in buildings. Energy Build 39:823–836. https://doi.org/10.1016/j.enbuild.2007.02.003

13. Andersen RK, Fabi V, Corgnati SP (2016) Predicted and actual indoor environmental quality: verification of occupants' behaviour models in residential buildings. Energy Build 127: 105–115. https://doi.org/10.1016/j.enbuild.2016.05.074

14. Fabi V, Andersen RV, Corgnati SP (2013) Influence of occupant's heating set-point preferences on indoor environmental quality and heating demand in residential buildings. HVAC&R Res 19:37–41. https://doi.org/10.1080/10789669.2013.789372

15. D'Oca S, Fabi V, Corgnati SP, Andersen RK (2014) Effect of thermostat and window opening occupant behavior models on energy use in homes. Build Simul 7:683–694. https://doi.org/ 10.1007/s12273-014-0191-6

16. Santin OG (2011) Behavioural patterns and user profiles related to energy consumption for heating. Energy Build 43:2662–2672. https://doi.org/10.1016/j.enbuild.2011.06.024

17. Gunay HB, Brien WO, Beausoleil-Morrison I (2016) A toolkit for developing data-driven occupant behaviour and presence models. In: eSim. p paper 58

18. Yan D, O'brien W, Hong T et al (2015) Occupant behavior modeling for building performance simulation: current state and future challenges. Energy Build 107:264–278. https://doi.org/10.1016/j.enbuild.2015.08.032

19. Haldi F, Robinson D (2011) The impact of occupants' behaviour on building energy demand. J Build Perform Simul 4:323–338. https://doi.org/10.1080/19401493.2011.558213

20. Wang C, Yan D, Jiang Y (2011) A novel approach for building occupancy simulation. Build Simul 4:149–167. https://doi.org/10.1007/s12273-011-0044-5

21. Widén J, Nilsson AM, Wäckelgård E (2009) A combined Markov-chain and bottom-up approach to modelling of domestic lighting demand. Energy Build 41:1001–1012. https://doi. org/10.1016/j.enbuild.2009.05.002

22. Gunay HB, O'Brien W, Beausoleil-Morrison I et al (2014) Coupling stochastic occupant models to building performance simulation using the discrete event system specification formalism. J Build Perform Simul 7:457–478. https://doi.org/10.1080/19401493.2013.866695

23. Wang D, Federspiel CC, Rubinstein F (2005) Modeling occupancy in single person offices. Energy Build 37:121–126. https://doi.org/10.1016/j.enbuild.2004.06.015
24. Gunay HB, O'Brien W, Beausoleil-Morrison I (2013) A critical review of observation studies, modeling, and simulation of adaptive occupant behaviors in offices. Build Environ 70:31–47. https://doi.org/10.1016/j.buildenv.2013.07.020
25. Hong T, Chen Y, Belafi Z, D'Oca S (2017) Occupant behavior models: a critical review of implementation and representation approaches in building performance simulation programs. Build Simul. https://doi.org/10.1007/s12273-017-0396-6
26. Feng X, Yan D, Wang C (2016) On the simulation repetition and temporal discretization of stochastic occupant behaviour models in building performance simulation. J Build Perform Simul 1–13. https://doi.org/10.1080/19401493.2016.1236838
27. Warren P, Parkins L (1984) Window-opening behaviour in office buildings. Build Serv Eng Res Technol 5:89–101
28. Johnson T, Long T (2005) Determining the frequency of open windows in residences: a pilot study in Durham, North Carolina during varying temperature conditions. J Expo Anal Environ Epidemiol 15:329–349. https://doi.org/10.1038/sj.jea.7500409
29. Li N, Li J, Fan R, Jia H (2015) Probability of occupant operation of windows during transition seasons in office buildings. Renew Energy 73:84–91. https://doi.org/10.1016/j.renene.2014.05.065
30. Mahdavi A, Tahmasebi F, Kayalar M (2016) Prediction of plug loads in office buildings: simplified and probabilistic methods. Energy Build 129:322–329. https://doi.org/10.1016/j.enbuild.2016.08.022
31. Wang C, Yan D, Sun H, Jiang Y (2016) A generalized probabilistic formula relating occupant behavior to environmental conditions. Build Environ 95:53–62. https://doi.org/10.1016/j.buildenv.2015.09.004
32. Ren X, Yan D, Wang C (2014) Air-conditioning usage conditional probability model for residential buildings. Build Environ 81:172–182. https://doi.org/10.1016/j.buildenv.2014.06.022
33. Hong T, Sun H, Chen Y et al (2015) An occupant behavior modeling tool for co-simulation. Energy Build 117:272–281. https://doi.org/10.1016/j.enbuild.2015.10.033
34. Herkel S, Knapp U, Pfafferott J (2008) Towards a model of user behaviour regarding the manual control of windows in office buildings. Build Environ 43:588–600. https://doi.org/10.1016/j.buildenv.2006.06.031
35. Zhang Y, Barrett P (2012) Factors influencing the occupants' window opening behaviour in a naturally ventilated office building. Build Environ 50:125–134. https://doi.org/10.1016/j.buildenv.2011.10.018
36. Sadeghi SA, Awalgaonkar NM, Karava P, Bilionis I (2017) A Bayesian modeling approach of human interactions with shading and electric lighting systems in private offices. Energy Build 134:185–201. https://doi.org/10.1016/j.enbuild.2016.10.046
37. Gaetani I, Hoes PJ, Hensen JLM (2016) Occupant behavior in building energy simulation: towards a fit-for-purpose modeling strategy. Energy Build 121:188–204. https://doi.org/10.1016/j.enbuild.2016.03.038
38. Schakib-Ekbatan K, Çakıcı FZ, Schweiker M, Wagner A (2015) Does the occupant behavior match the energy concept of the building?—Analysis of a German naturally ventilated office building. Build Environ 84:142–150. https://doi.org/10.1016/j.buildenv.2014.10.018
39. Lee YS, Malkawi AM (2014) Simulating multiple occupant behaviors in buildings: an agent-based modeling approach. Energy Build 69:407–416. https://doi.org/10.1016/j.enbuild.2013.11.020
40. Fanger PO (1970) Thermal comfort: analysis and applications in environmental engineering. Mc Graw-Hill, New York
41. Powers WT (1973) Behavior: the control of perception. https://doi.org/10.2307/2066319
42. Zhao X, Venkateswaran J, Son Y (2005) Modeling human operator decision-making in manufacturing systems using BDI agent paradigm. In: Annual industrial engineering research conference. pp 14–18

43. Langevin J, Wen J, Gurian PL (2014) Simulating the human-building interaction: Development and validation of an agent-based model of office occupant behaviors. Build Environ 88:27–45. https://doi.org/10.1016/j.buildenv.2014.11.037
44. Stazi F, Naspi F, D'Orazio M (2017) A literature review on driving factors and contextual events influencing occupants' behaviours in buildings. Build Environ 118:40–66. https://doi.org/10.1016/j.buildenv.2017.03.021
45. Yang J, Santamouris M, Lee SE (2016) Review of occupancy sensing systems and occupancy modeling methodologies for the application in institutional buildings. Energy Build 121: 344–349. https://doi.org/10.1016/j.enbuild.2015.12.019
46. Oldewurtel F, Sturzenegger D, Morari M (2013) Importance of occupancy information for building climate control. Appl Energy 101:521–532. https://doi.org/10.1016/j.apenergy.2012.06.014
47. Chang WK, Hong T (2013) Statistical analysis and modeling of occupancy patterns in open-plan offices using measured lighting-switch data. Build Simul 6:23–32. https://doi.org/10.1007/s12273-013-0106-y
48. Dodier RH, Henze GP, Tiller DK, Guo X (2006) Building occupancy detection through sensor belief networks. Energy Build 38:1033–1043. https://doi.org/10.1016/j.enbuild.2005.12.001
49. Page J, Robinson D, Morel N, Scartezzini JL (2008) A generalised stochastic model for the simulation of occupant presence. Energy Build 40:83–98. https://doi.org/10.1016/j.enbuild.2007.01.018
50. Mahdavi A, Tahmasebi F (2015) Predicting people's presence in buildings: an empirically based model performance analysis. Energy Build 86:349–355. https://doi.org/10.1016/j.enbuild.2014.10.027
51. D'Oca S, Hong T (2015) Occupancy schedules learning process through a data mining framework. Energy Build 88:395–408. https://doi.org/10.1016/j.enbuild.2014.11.065
52. Chen Z, Soh YC (2016) Comparing occupancy models and data mining approaches for regular occupancy prediction in commercial buildings. J Build Perform Simul 1493:1–9. https://doi.org/10.1080/19401493.2016.1199735
53. Coakley D, Raftery P, Keane M (2014) A review of methods to match building energy simulation models to measured data. Renew Sustain Energy Rev 37:123–141. https://doi.org/10.1016/j.rser.2014.05.007
54. Chen Z, Masood MK, Soh YC (2016) A fusion framework for occupancy estimation in office buildings based on environmental sensor data. Energy Build 133:790–798. https://doi.org/10.1016/j.enbuild.2016.10.030
55. Liao C, Lin Y, Barooah P (2012) Agent-based and graphical modelling of building occupancy. J Build Perform Simul 5:5–25. https://doi.org/10.1080/19401493.2010.531143
56. Feng X, Yan D, Hong T (2015) Simulation of occupancy in buildings. Energy Build 87: 348–359. https://doi.org/10.1016/j.enbuild.2014.11.067
57. Luo X, Lam KP, Chen Y, Hong T (2017) Performance evaluation of an agent-based occupancy simulation model. Build Environ 115:42–53. https://doi.org/10.1016/j.buildenv.2017.01.015
58. Menezes AC, Cripps A, Bouchlaghem D, Buswell R (2012) Predicted vs. actual energy performance of non-domestic buildings: using post-occupancy evaluation data to reduce the performance gap. Appl Energy 97:355–364. https://doi.org/10.1016/j.apenergy.2011.11.075
59. Cowie A, Hong T, Feng X, Darakdjian Q (2017) Usefulness of the obFMU module examined through a review of occupant modelling functionality in building performance simulation programs. In: IBPSA Building Simulation Conference, San Francisco, USA
60. de Dear RJ, Brager GS (2002) Thermal comfort in naturally ventilated buildings: revisions to ASHRAE Standard 55. Energy Build 34:549–561. https://doi.org/10.1016/S0378-7788(02)00005-1
61. Wang L, Greenberg S (2015) Window operation and impacts on building energy consumption. Energy Build 92:313–321. https://doi.org/10.1016/j.enbuild.2015.01.060

62. Peter G, Paul A, Drury B (2007) Simulation of energy management systems in EnergyPlus. In: Building Simulation, Beijing, China. pp 1–9

63. Hong T, Taylor-Lange SC, D'Oca S et al (2016) Advances in research and applications of energy-related occupant behavior in buildings. Energy Build 116:694–702. https://doi.org/10.1016/j.enbuild.2015.11.052

64. Noidui TS, Wetter M, Zuo W (2013) Functional mock-up unit import in EnergyPlus for co-simulation. In: Conference International Building Performing Simulation Association Chambery, France. pp 3275–3282

65. Yao J, Chow DHC, Zheng R-Y, Yan C-W (2015) Occupants' impact on indoor thermal comfort: a co-simulation study on stochastic control of solar shades. J Build Perform Simul 1493:1–16. https://doi.org/10.1080/19401493.2015.1046492

66. Langevin J, Wen J, Gurian PL (2016) Quantifying the human-building interaction: considering the active, adaptive occupant in building performance simulation. Energy Build 117:372–386. https://doi.org/10.1016/j.enbuild.2015.09.026

67. Chen Y, Gu L, Zhang J (2015) EnergyPlus and CHAMPS-Multizone co-simulation for energy and indoor air quality analysis. Build Simul 8:371–380. https://doi.org/10.1007/s12273-015-0211-1

68. Kashif A, Ploix S, Dugdale J, Le XHB (2013) Simulating the dynamics of occupant behaviour for power management in residential buildings. Energy Build 56:85–93. https://doi.org/10.1016/j.enbuild.2012.09.042

69. IEA EBC (2013) Annex 66: definition and simulation of occupant behavior in buildings

70. Hong T, D'Oca S, Turner WJN, Taylor-Lange SC (2015) An ontology to represent energy-related occupant behavior in buildings. Part I: Introduction to the DNAs framework. Build Environ 92:764–777. https://doi.org/10.1016/j.buildenv.2015.02.019

71. Hong T, D'Oca S, Taylor-Lange SC et al (2015) An ontology to represent energy-related occupant behavior in buildings. Part II: Implementation of the DNAS framework using an XML schema. Build Environ 94:196–205. https://doi.org/10.1016/j.buildenv.2015.08.006

72. Belafi Z, Hong T (2016) A library of building occupant behavior models represented in a standardized schema. BEHAVE 2016 4th European Conference Behaviour Energy Efficency. pp 8–9

73. Belafi Z, Hong T, Reith A (2017) Smart building management vs. intuitive human control: lessons learnt from an office building in Hungary. Build. https://doi.org/10.1007/s12273-017-0361-4

74. Haldi F, Robinson D (2011) Modelling occupants' personal characteristics for thermal comfort prediction. Int J Biometeorol 55:681–694. https://doi.org/10.1007/s00484-010-0383-4

75. Schweiker M, Haldi F, Shukuya M, Robinson D (2012) Verification of stochastic models of window opening behaviour for residential buildings. J Build Perform Simul 5:55–74. https://doi.org/10.1080/19401493.2011.567422

76. Nicol JF (2001) Characterising occupant behavior in buildings: towards a stochastic model of occupant use of windows, lights, blinds heaters and fans. In: Seventh International IBPSA Conference. pp 1073–1078

77. James G, Witten D, Hastie T, Tibshirani R (2013) An introduction to statistical learning. Curr Med Chem. https://doi.org/10.1007/978-1-4614-7138-7

78. Fritsch R, Kohler A, Nygård-Ferguson M, Scartezzini J-L (1990) A stochastic model of user behaviour regarding ventilation. Build Environ 25:173–181. https://doi.org/10.1016/0360-1323(90)90030-U

79. Fabi V, Andersen RK, Corgnati S (2015) Verification of stochastic behavioural models of occupants' interactions with windows in residential buildings. Build Environ 94:371–383. https://doi.org/10.1016/j.buildenv.2015.08.016

Chapter 8
Conclusions and Future Challenges

Abstract Occupants' behaviour in buildings has been recently recognised as one of the main aspects influencing the overall energy consumptions. The users' decision-making process is a complex mechanism, affected by surrounding's conditions, personal features and random events. However, people behaviour is usually underestimated since, in the simulation environment, it is reproduced according to pre-determined and fixed rules. This static and oversimplified approach led to incorrect evaluations, energy wastes and big discrepancies between real and predicted building performance. Such considerations are even more relevant for the nZEBs. Due to the recent regulations on the energy reduction, the nZEBs' diffusion is (and will be) the main target of the building sector. As a consequence, reliable evaluations are essential to reach the nZEB target. In this view, this book aims at offering a complete overview of the human perspective, especially in nZEBs. In particular, this final Chapter aims at providing recommendations and suggestions for a correct inclusion of the behavioural components in the building process and it also reports some open problems and future directions. The considerations regard many different (but strictly connected) aspects: the triggers of users' behaviour, the effects of users' actions both in environmental and energy terms, the methods to acquire behavioural data and the development of behavioural models.

8.1 Main Conclusions

Combining behavioural evaluations to the latest technologies is crucial to match the nZEB requirements. Erroneous assumptions on users' behaviours at the design, simulation and operation phases represent the greatest part of the performance gap. In fact, many post-occupancy analyses highlighted that nZEBs often perform worse than predicted and also worse than their standard counterpart. The performance gap can be bridged only with a complete understanding and a correct reproduction of occupants' behaviours. Such aspects have been investigated, studying both the environment and the subjects and trying to understand their mutual effect and interactions.

© The Author(s) 2018   79
F. Stazi and F. Naspi, *Impact of Occupants' Behaviour on Zero-Energy Buildings*,
SpringerBriefs in Energy, https://doi.org/10.1007/978-3-319-71867-5_8

Studies on the environmental trends highlighted that people presence and behaviours have an undeniable effect on the indoor environment. Moreover, their influence changes according to some specific features. For example, the impact seems to be greater during transition seasons (i.e. spring and autumn) due to the wide variability of environmental conditions. Also the occupants' density (connected to the building use) affects the indoor environment: the more the density increases (e.g. school classrooms) the more the users perturb the environment.

To predict how much the occupants will impact on the building environment and consumptions, many surveys aimed at understanding the drivers for users' behaviours. Occupants' actions are triggered by many different stimuli, related to both objective and subjective aspects. The former, which can be identified using sensors and surveys, concerns environmental (e.g. temperature), time-related (e.g. entering the room) and contextual factors (e.g. interior design). The latter are connected to features that are peculiar of each person (i.e. physiological, psychological and social factors). The quantification of subjective features is more difficult in comparison to objective ones since an intrusion in the occupants' life is unavoidable. Moreover, several studies highlighted that people's behaviour can change in low-energy buildings both in a positive and negative way. These findings underline the necessity of a multidisciplinary approach, which can link knowledge from different fields, bringing to a better understanding of the human nature.

Interactions on building systems and devices have a greater impact than personal modifications. The adaptive measures, taken to restore the comfort, have a direct or indirect influence on both energy consumptions and users' sensation. In particular, adjustments on heating and cooling systems are the principal influencing factors, especially in residential buildings. However, also the demand for lighting and ventilation plays an undeniable role. Some devices, as fans and doors, have a little direct effect on building performance but they can significantly affect people's perceptions. Users can also have different choices to reach the same goal and the sequence of their action is another key aspect for the building optimisation. Unfortunately, studies on this aspect are still few, mainly directed to office buildings and proposing different sequences. In general, in offices people tend to adjust, at first, at the personal level (e.g. clothing) and then at room level (e.g. window opening). Similar trends have been also observed in nZEBs but only few studies are available till now.

The experimental acquisition of occupancy and behavioural data is essential for the development of behavioural models which predict the human-building interaction and to optimise intelligent building controls. Sensor networks allow a simultaneous recording of several actions, while Wi-Fi based technology proved to be extremely useful to reduce cabling and intrusiveness while providing applications at large scales. The diffusion of Building Management Systems is seen as a valid aid for data collection, post-occupancy evaluations and real-time adjustments.

The subsequent step to optimise building design and management (also through intelligent and adaptive devices) concerns the development of behavioural models. The modelling approaches to predict both users' presence and behaviours are becoming more and more accurate. They are moving towards stochastic approaches and especially ABMs which better represent human randomness. However, to reach reliable simulations' results and match real users' preferences, a more exhaustive knowledge of the human component is essential.

8.2 Future Challenges and Open Issues

Despite big progress have been reached in the investigation of the human-building interaction, many challenges must be faced jet.

Till now, the most of the researchers studied a limited portion of building uses and many climate areas remain unexplored. Moreover, they focused mainly on objective aspects since they are easy to record and analyse. However, having a comprehensive overview of occupants' behaviour is extremely important to definitely identify triggering factors and their weight in different contexts. Investigations in specific climate conditions (e.g. Mediterranean area), building uses (e.g. schools) as well as in relation to subjective features (e.g. background, social and group phenomena) are deeply recommended.

Efforts should be made also to include advanced occupant modelling features in BEPS. Even if this will require further user inputs, it is seen as an essential step to improve the accuracy of simulated results [1]. In parallel, the adoption of robust methods and standard process [2, 3], will also aid researchers and designers to develop models and improve comparisons between different contexts.

Some further considerations are needed for the particular conditions of nZEBs. Although their diffusion is increasing, studies on occupants' behaviours in such contexts are still limited and they report both negative and positive influences of users' adaptation. Further investigations are needed to understand the key aspects which lead to enhancement of people's behaviours, so by promote them.

Nowadays, the nZEB target has been reached relying only on technological and physical features. However, the discrepancies between real and predicted performance highlighted that counting only on technology is extremely difficult to drive down energy use [2]. In fact, the nZEB requirements cannot be achieved without considering occupants' impact on the building performance along all the building life cycle [4]. So, the understanding of the users' decision-making process is essential to calibrate building automation systems which seem to be crucial in nZEBs [5].

The findings derived from many studies highlighted some useful strategies for building energy reduction. One way concerns the reduction of users' impact on building performance, making the buildings less sensitive to occupants' behaviours [6]. The opposite way is to focus on the people through encouragements in changing their behaviours and through education of building use [7]. The synergy of both these approaches would probably be the best solution since it could bring to sensible energy savings [5, 6].

This book aimed at underling the importance of the human dimension and the usefulness of its inclusion in all the phases of the building process. Following a behavioural approach, enhancements are expected for designers, end-users and building's operators. The provided findings could guide towards strategies and interventions to concretely reduce the energy use and improve the users' comfort. This approach can also be adapted to other contexts concerning the behavioural sphere and the interaction between the users and the built environment, such as human safety, security and space fruition.

References

1. O'Brien W, Gaetani I, Gilani S, et al (2016) International survey on current occupant modelling approaches in building performance simulation. J Build Perform Simul 1–19. https://doi.org/10.1080/19401493.2016.1243731
2. Hong T, D'Oca S, Turner WJN, Taylor-Lange SC (2015) An ontology to represent energy-related occupant behavior in buildings. Part I: Introduction to the DNAs framework. Build Environ 92:764–777. https://doi.org/10.1016/j.buildenv.2015.02.019
3. Hong T, D'Oca S, Taylor-Lange SC et al (2015) An ontology to represent energy-related occupant behavior in buildings. Part II: Implementation of the DNAS framework using an XML schema. Build Environ 94:196–205. https://doi.org/10.1016/j.buildenv.2015.08.006
4. D'Oca S, Hong T, Langevin J (2018) The human dimensions of energy use in buildings: a review. Renew Sustain Energy Rev 81:731–742. https://doi.org/10.1016/j.rser.2017.08.019
5. Fabi V, Spigliantini G, Corgnati SP (2017) Insights on smart home concept and occupants' interaction with building controls. Energy Procedia 111:759–769. https://doi.org/10.1016/j.egypro.2017.03.238
6. Karjalainen S (2016) Should we design buildings that are less sensitive to occupant behaviour? A simulation study of effects of behaviour and design on office energy consumption. Energy Effic 9:1257–1270. https://doi.org/10.1007/s12053-015-9422-7
7. Belafi Z, Hong T, Reith A (2017) Smart building management vs. intuitive human control: Lessons learnt from an office building in Hungary. Build Simul https://doi.org/10.1007/s12273-017-0361-4

Appendix A
Experimental Cases Study

Case Study A: Office Building

The case study is a university building, mainly composed of classrooms and offices (Fig. A.1). It is settled at Ancona (central Italy) which is characterised by a Mediterranean climate (Table A.1). The construction, built in the Seventies,

Fig. A.1 Office building external view

Table A.1 Main features of the office building

Building features	
Location	Ancona, Italy
Latitude	43° 35′ 12″ 40 N
Longitude	13° 30′ 59″ 74 E
Altitude	140 m
Climate (Köppen classification)	Hot-summer Mediterranean climate

© The Author(s) 2018
F. Stazi and F. Naspi, *Impact of Occupants' Behaviour on Zero-Energy Buildings*,
SpringerBriefs in Energy, https://doi.org/10.1007/978-3-319-71867-5

presents a concrete load-bearing structure and it is covered by ribbon windows in aluminium frame and double glazing.

Three offices have been selected for the experimental study (Fig. A.2). They are placed on the same floor level and present similar features. Two adjacent offices (A and B) are north-oriented, while the third room (C) faces east. Two or three persons occupy the rooms from Monday to Friday and from 9 a.m. to 7 p.m (Table A.2).

Fig. A.2 The surveyed offices

Table A.2 Main features of the offices

Rooms features								
Room	Net floor area (m²)	Internal Height (m)	Heated volume (m³)	Ratio S/V	Orientation	Number of persons	Glazed surface (m²)	Opening surface (m²)
A	20	3	60	0.33	North	3	6	3
B	20	3	60	0.33	North	3	6	3
C	15	3	45	0.33	East	2	6	3

The monitoring lasted one year, allowing acquisitions from all the seasons. Each room has been equipped with the same sensors to acquire objective parameters (Fig. A.3). The subjects were also asked to fill out nameless questionnaires to achieve subjective sensations on the thermal environment (Tables A.3 and A.4).

Plant view

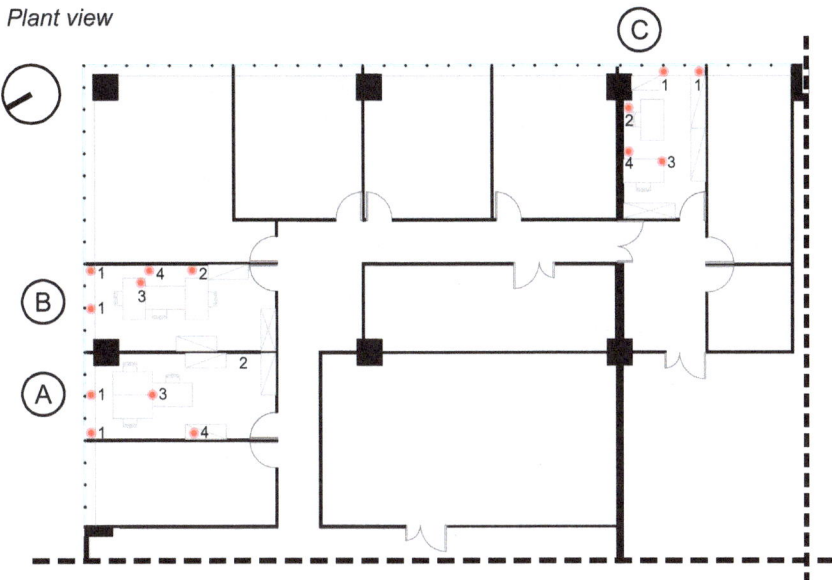

Acquisition systems   *Thermal vote from questionnaires*

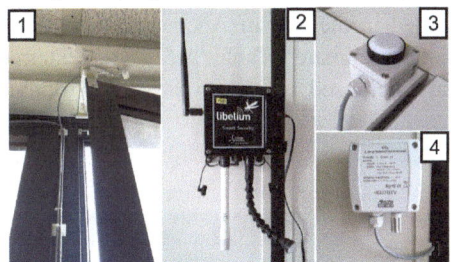   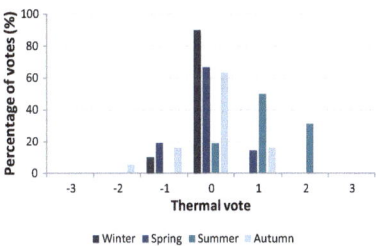

Legend

1)Windows status (boolean)
2)Indoor temperature (°C), indoor RH (%), occupancy
3)Work-plane illuminance (lx)
4)CO_2 concentration (ppm)

Fig. A.3 Plant view with indication of the sensors' position; Identification of the installed instruments; Thermal comfort trend from the questionnaires

Table A.3 Features of the monitoring in the offices

Monitoring features	
Length	1 year
Duration	May 2016–May 2017
Number of the subjects involved	8 persons (3 male and 5 female)
Age of the subjects	Between 27 and 34 years (mean: 30)

Table A.4 Main characteristics of the probes installed in the offices

Probes features				
Parameter	Sensor	Accuracy	Range	Number
Air temperature (°C)	Thermistor (SHT75)	±0.4 °C	0–70 °C	3 (1 per room)
Relative Humidity (%)	Capacitive (SHT75)	±1.8%	0–100%	3 (1 per room)
CO_2 (ppm)	NDIR	±50 ppm	0–2000 ppm	3 (1 per room)
Light (lux)	Photodiode Si	±3%	0.02–20 klx	3 (1 per room)
Occupancy (num. occupants)	PIR	n/a	12 m	3 (1 per room)
Windows Openings (num.)	Magnetic	n/a	n/a	12 (4 per room)

Case Study B: School Building

The case study is a school complex (Fig. A.4). It is composed of three main blocks which have different purposes (classrooms and laboratories, offices, auditorium and gym). The institute, built in 2010, is settled in the city of Ancona (central Italy) which is characterised by a Mediterranean climate (Table A.5). The external envelope presents cavity walls with an interposed insulation layer and double glazing windows in aluminium frame.

Fig. A.4 School external view

Table A.5 Main features of the school building

Building features	
Location	Ancona, Italy
Latitude	43° 58′ 49″ 30 N
Longitude	13° 52′ 57″ 01 E
Altitude	67 m
Climate (Köppen classification)	Hot-summer Mediterranean climate

The experimental study concerns two adjacent classrooms (Fig. A.5). Both of them are placed on the ground floor and present very similar features, allowing a comparison between the two rooms (Table A.6). The students attend lessons from Monday to Friday and from 8 a.m. to 2 p.m. Class A has been equipped with a motorised system for window opening to optimise the indoor environmental quality, while in class B the windows were operated manually (for further details refer to Appendix B).

Fig. A.5 The surveyed classrooms

Table A.6 Main features of the classrooms

Rooms features								
Room	Net floor area (m²)	Internal Height (m)	Heated volume (m³)	Ratio S/V	Orientation	Number of persons	Glazed surface (m²)	Opening surface (m²)
A	53	3.2	169	0.31	North	17	6	6
B	55	3.2	175	0.31	North	16	6	6

The monitoring started in class B (i.e. the manual) to investigate thermal environment and occupants' behaviours. Then, it continued in parallel in both the classrooms (Fig. A.6). The same environmental parameters and actions on windows have been recorded in the two rooms, allowing a comparison between the two managements. Students and teachers were also asked to fill out nameless questionnaires to achieve subjective perceptions on thermal environment and indoor air quality (Tables A.7 and A.8).

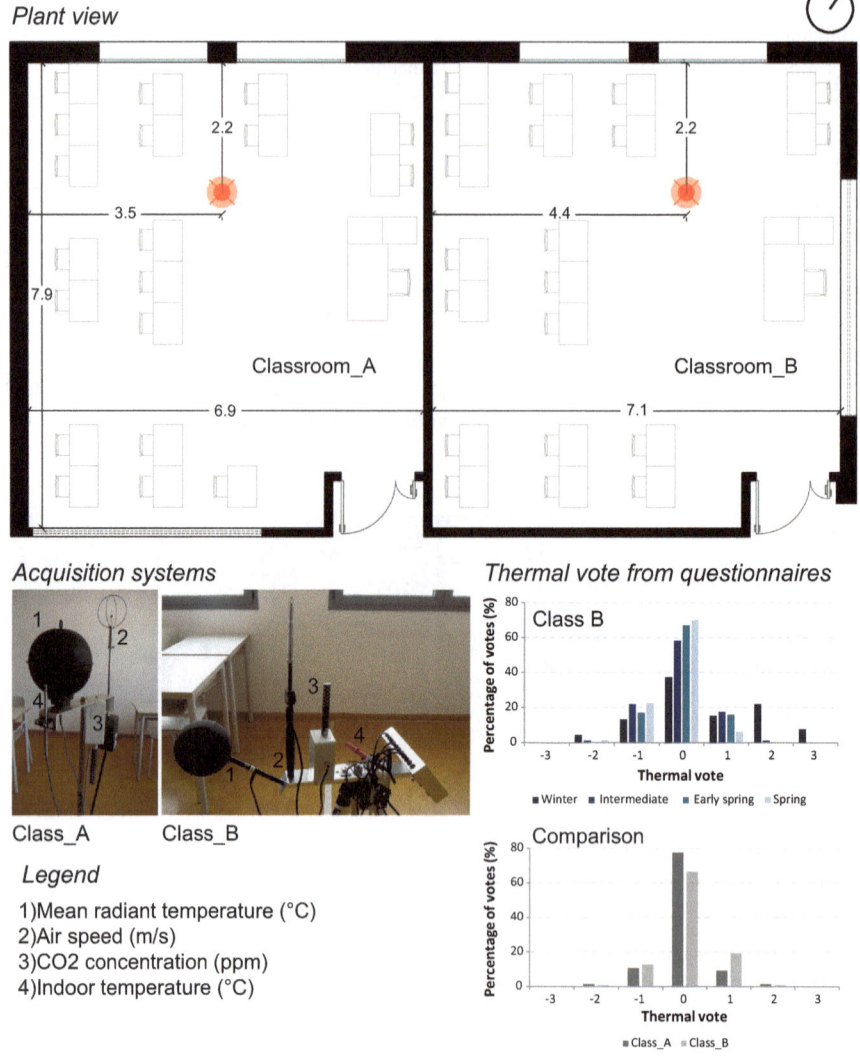

Fig. A6 Plant view with indication of the sensors' position; Identification of the installed instruments; Thermal comfort votes from the questionnaires

Table A.7 Features of the monitoring in the classrooms

Monitoring features	
Length	56 days
Duration	15–28 January 2013/19th March–29th April 2015
Number of the subjects involved	33 persons (5 male and 28 female)
Age of the subjects	14–15 years old

Table A.8 Main characteristics of the probes installed in the classrooms

Probes features				
Parameter	Sensor	Accuracy	Range	Number
Air temperature (°C)	Thermistor (PT100)	± 0.15 °C	−50 to 80 °C	2 (1 per room)
Mean radiant temperature (°C)	Globe thermometer	± 0.15 °C	−40 to 80 °C	2 (1 per room)
CO_2 concentration (ppm)	IR	20–70 ppm	0–3000 ppm	2 (1 per room)
Air speed (m/s)	Hot sphere anemometer	±0.01 m/s	0.05–1 m/s	1 (Class A)
Air speed (m/s)	Hot wire anemometer	0–0.5 m/s	0.01–20 m/s	1 (Class B)

Case Study C: Residential Building

The case study is a residential building built in the early 1900s and recently renovated (Fig. A.7). The house, occupied by a family of four persons, is settled in the city of Cattolica (central Italy) which is characterised by a Mediterranean climate (Table A.9). It is a two floors masonry building with an insulation layer and double glazing windows (Table A.10).

Fig. A.7 Residential building external views

Table A.9 Main features of the residential building

Building features	
Location	Cattolica, Italy
Latitude	43° 57′ 55.5″ N
Longitude	12° 44′ 36.8″ E
Altitude	5 m
Climate (Köppen classification)	Hot-summer Mediterranean climate

Table A.10 Main features of the house

House features						
Net floor area (m²)	Internal Height (m)	Heated volume (m³)	Ratio S/V	Number of persons	Glazed surface (m²)	Opening surface (m²)
115	3	385	0.74	4	13	13

The monitoring allowed acquisitions from different seasons (i.e. summer, autumn and winter). The sensors to record indoor and outdoor parameters have been placed in a room on the second floor (Fig. A.8, Tables A.11 and A.12).

Plant view

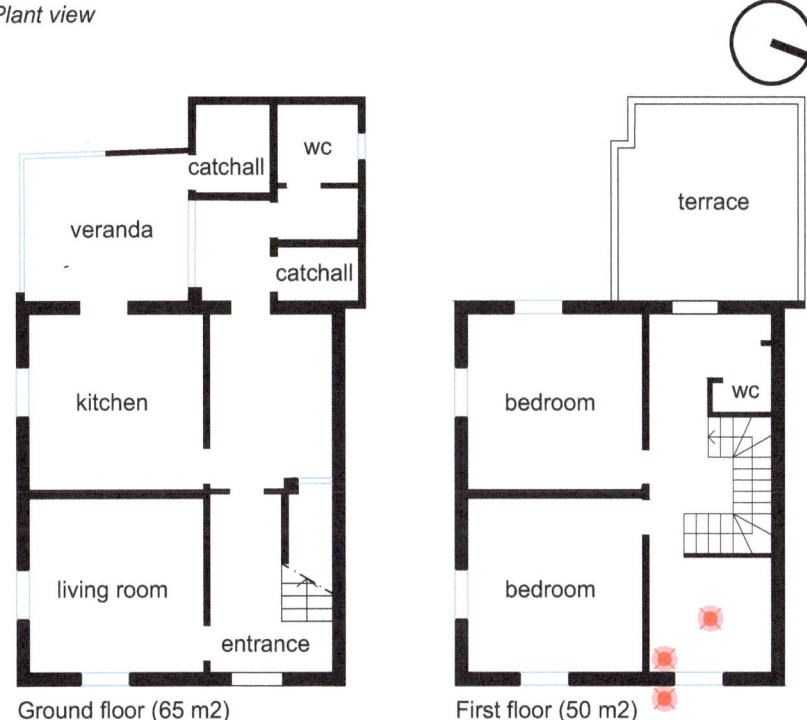

Ground floor (65 m2)   First floor (50 m2)

Acquisiton system

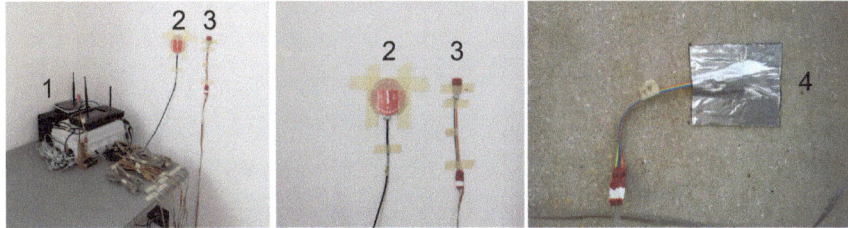

Legend
1)Data logger
2)Heat flux (W/m2K)
3)Surface indoor temperature (°C) & humidity (%)
4)Surface outdoor temperature (°C) & humidity (%)

Fig. A.8 Plant view with indication of the sensors' position and identification of the installed instruments

Table A.11 Features of the monitoring in the house

Monitoring features	
Length	130 days
Duration	1st August–31th October 2015 and 14th January–22th February 2017
Number of the subjects involved	4 persons (2 male and 2 female)
Age of the subjects	Between 17 and 52 years old

Table A.12 Main characteristics of the probes installed in the house

Probes features				
Parameter	Sensor	Accuracy	Range	Number
Outdoor air temperature (°C)	Digital sensor	±0.3 °C	−40 to 125 °C	1
Outdoor humidity (%)		±2%	0–100%	1
Indoor air temperature (°C)	Digital sensor	± 0.3 °C	−40 to 125 °C	2
Indoor humidity (%)		±2%	0–100%	2
Heat flux (W/m^2 K)	Thermopile	±5%	−30 to 70 °C	1

Appendix B
Examples of Applications

An automatic system, driven by an adaptive control algorithm, has been installed in a school classroom to optimise indoor environmental quality, especially the indoor temperature and the CO_2 levels (Fig. B.1 and Table B.1). The workflow of the automatic system is presented in Fig. B.2 and specified in Table B.2. The classroom indoor environment was compared to that of an adjacent class to assess the enhancement of the system (Fig. B.3). The specific features of the classrooms and the sensors were reported in Appendix A case study B.

Fig. B.1 Internal view of the classroom

Table B.1 Main features of the classroom

Room features								
Room	Net floor area (m²)	Internal Height (m)	Heated volume (m³)	Ratio S/V	Orientation	Number of persons	Glazed surface (m²)	Opening surface (m²)
A	53	3.2	169	0.31	North	17	6	6

© The Author(s) 2018
F. Stazi and F. Naspi, *Impact of Occupants' Behaviour on Zero-Energy Buildings*,
SpringerBriefs in Energy, https://doi.org/10.1007/978-3-319-71867-5

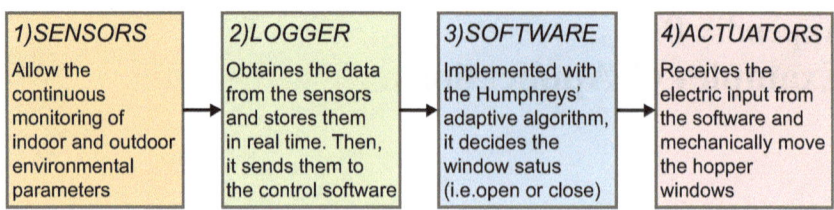

Fig. B.2 Schema of the automatic system

Table B.2 Main element of the system

Element	Features
(1) Sensors	They acquired several environmental parameters (i.e. indoor temperature, mean radiant temperature, air speed and CO_2 concentration) with a rate of 1 min
(2) Datalogger	The logger, directly connected to the sensors, stored the data in their own memory
(3) Software	The control algorithm was developed using the graphical software LabVIEW 2014 The Humphreys' adaptive algorithm was adopted but slightly modified At each time step, the algorithm made a double comparison. It compared the actual indoor temperature to a comfort range (Comfort temperature ± 1 °C) and the actual CO_2 concentration to an imposed limit (1250 ppm) The window status depended on the results of these comparisons and on the previous window state (e.g. if the window was closed and the CO_2 concentration was higher than 1250 ppm, the window was open)
(4) Actuators	The instruments were power supply (24 V) and equipped with a metal chain of 25 cm Installed on the top of the windows, they moved them in parallel A neoprene cuff was placed on the actuators to reduce the bother during the lessons

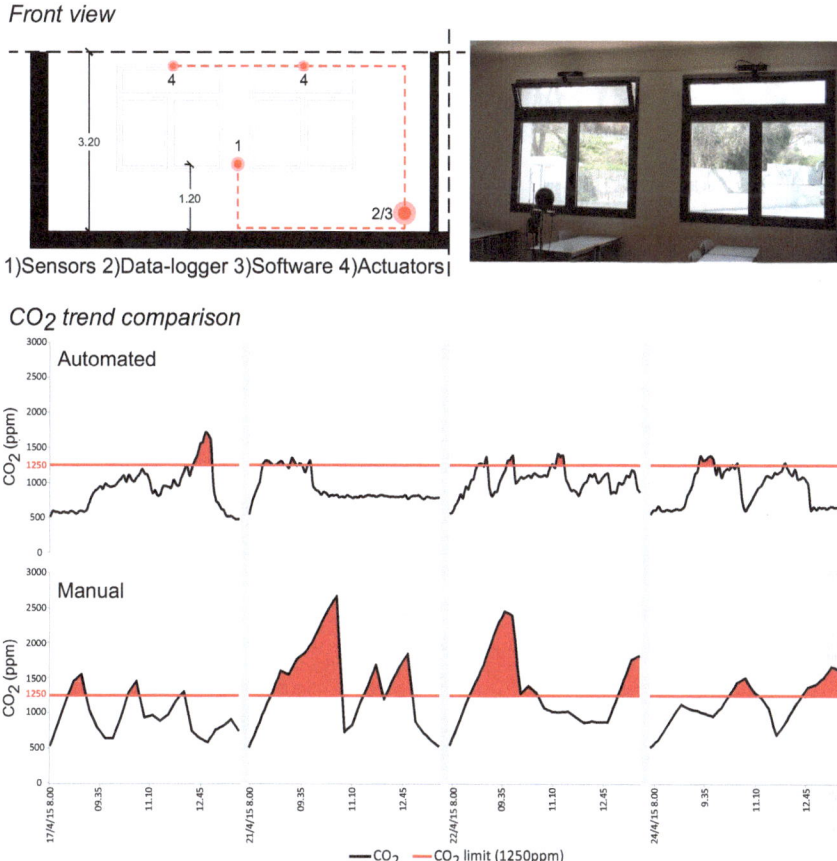

Fig. B.3 Front view with indication of the sensors and actuators' position; comparison of the CO_2 trends between the room equipped with the automatic system and one manually operated